SpringerBriefs in History of Science
and Technology

More information about this series at http://www.springer.com/series/10085

Massimiliano Badino

The Bumpy Road

Max Plank from Radiation Theory
to the Quantum (1896–1906)

Massimiliano Badino
Universitat Autònoma de Barcelona
Barcelona
Spain

ISSN 2211-4564   ISSN 2211-4572   (electronic)
SpringerBriefs in History of Science and Technology
ISBN 978-3-319-20030-9   ISBN 978-3-319-20031-6   (eBook)
DOI 10.1007/978-3-319-20031-6

Library of Congress Control Number: 2015941350

Springer Cham Heidelberg New York Dordrecht London

Printed on acid-free paper

Springer International Publishing AG Switzerland is part of Springer Science+Business Media
(www.springer.com)

To Emma
running, not crawling

Preface

The birth of quantum theory is one of the most impressive examples of scientific revolution. What makes this historical episode fascinating is that the breakthrough originated from a 'classical' research program that progressively accumulated conceptual and experimental tensions up to the point of generating a 'non-classical' outcome. The progressiveness, however, does not make the historical reconstruction any simpler. The tight interplay of highly technical issues belonging to fields as diverse as electromagnetism, thermodynamics, and combinatorics is a serious hindrance for many historians to the extent that a full account of this interplay remains a desideratum.

Such an account is not solely interesting for the historian of physics. Indeed, what follows can be viewed as a study in the epistemological problem of theory assemblage. Traditional epistemology tends to regard a scientific theory as a given that can be confirmed, falsified, modified, or overthrown by a new one. But very little has been hitherto said on the process that brings a theory into being. The case of quantum theory gives us an almost unique opportunity to investigate how progressively a new conception acquires a shape, how flexibly this shape is adapted to new insights, and, eventually, how dramatically it can change. For this reason, my approach is organized in order to spell out the morphology, the epistemic architecture, and the conceptual resources that form the building of Planck's theory. Only through a comparative analysis of these elements is it possible to understand how Planck's theory could embody the tensions at the border of different areas of knowledge.

Initially, Planck envisioned his theory as an integration between thermodynamics and electromagnetism. Within these two areas of knowledge, however, he was referring to specific theoretical sub-traditions that distinguished his program from those of his contemporaries. As the program unfolded, Planck had to take up new mathematical and conceptual techniques from kinetic theory and adapt them to his program. This is again an epistemologically interesting process that affects the structure of the theory and its capacity to respond to external challenges. Furthermore, the introduction of new resources and techniques does not automatically entail their integration into the program. Instead, Planck wanted to maintain the original bulk of his idea: to provide an explanation of the equilibrium within the cavity radiation that would stress the strict irreversibility of the equilibration process. The combination of Planck's explanatory constraints and the techniques he increasingly admitted made his theory less and less flexible, until, eventually, this process isolated some parts of Planck's theory from the rest of physical knowledge. I argue that the quantum, the revolutionary and ultimately uncalled-for upshot of Planck's theory, was initially in epistemic isolation within the web of his conceptual and mathematical resources. Only later, after the careful analysis of scholars such as Einstein, Ehrenfest, Lorentz, and Jeans, it acquired a revolutionary epistemic status.

This work is the achievement of my 15-year-long interest in Planck's radiation theory. It is the third book I dedicate to Planck and most likely the last one. I began this journey humbly benefitting from the advice of a great scholar, my Ph.D. supervisor Evandro Agazzi, and I finished it eagerly stealing from the wisdom of another great scholar, Jürgen Renn. As a matter of fact, Jürgen was the first and most indefatigable supporter of this project, at times even more indefatigable than I. The intermediate stations of this long trip have been punctuated by many other colleagues and friends whose thoughts, remarks, suggestions, questions, and criticisms, permeate these pages. Nadia Robotti imparted a passion for history of physics and encouraged patient work on the papers. Some of our afternoon discussions at the Physics Department of the University of Genoa still stick in my mind and always will. I had the luck and the honor to be part of the History of Quantum Project of the Max Planck Institute for the History of Science. I partook in the pioneering effort to write a different kind of history of the quantum revolution, and I am indebted to that exciting research environment more than words can convey. I thank Alexander Blum, Arianna Borrelli, Bretislav Friedrich, Clayton Gearhart, Dieter Hoffman, Shaul Katzir, Martin Jähner, Michel Janssen, Ed Jerkowitz, Christian Joas, Marta Jordi, Christoph Lehner, Daniela Monaldi, Jaume Navarro, and Arne Schirrmacher for all their contributions to my research. I also thank Allan Needell for our discussions on Planck's combinatorial procedure. Matteo Valleriani granted me his friendship, which is more valuable than any intellectual insight I can possibly obtain. My work has been supported by several institutions, among which I want to thank particularly the Library of the Max Planck Institute for the History of Science (Berlin), the Archive for the History of the Max Planck Society, the Hayden Library and the Library Storage Annex at the Massachusetts Institute of Technology, and the Centre d'Història de la Ciència of the Universitat Autònoma de Barcelona. I have discussed some of the ideas of this book at the conference *Continuity and Discontinuity in the Physical Sciences since the Enlightenment* (American Institute of Physics, July 2011), at the & HPS4 meeting (Athens, March 2012), and at the *Lectio Commandiniana* (Urbino, April 2013); I am indebted to the participants of those events for the interesting discussions that followed the lectures. Now that I have finished these pages, on a bitterly cold Sunday afternoon, I realize that this slim book is a rather inadequate representation of such a long intertwining of careers, personal relations, intellectual exchanges, enthusiasm, and passions. Once again, the journey was more important than the destination.

Cambridge, USA   Massimiliano Badino
February 2015

Acknowledgments

This work has been conceived within the framework of the History of Quantum Physics Project of the Max Planck Institute for the History of Science (Berlin). During that period, I have also been supported by Research Grant no. I-1054-112.4/2009 of the German-Israeli Foundation. The final stage of this research has been developed as part of my project "Order/Chaos: Genealogy of Two Concepts in the Culture of European Mathematical Physics" funded by the Marie Sklodowska-Curie Actions Grant no. PIOF-GA-2013-623526.

Contents

Chapter 1
In the Theoretician's Workshop: Notes for a Historical and Philosophical Analysis of Theories

Abstract This chapter sets the historiographical framework for the rest of the book. I claim that philosophy and historiography of theories are currently utterly separated and this situation is detrimental to both. The reason of this separation, I argue, is that philosophy of science has maintained an intellectualistic stance toward scientific theories. In the course of time, this attitude has crystallized in three theses or dogmas about what a theory is, how it works, and how we should approach it. These dogmas have hindered a reconciliation between the analysis of the internal structure of theories and their being historical-cultural objects. In the final part of the chapter, I propose my view. By systematically turning the dogmas upside down, I argue that we should use reflectively our theoretical knowledge to shed light on theories as knowledge-production devices. I distinguish between the representational, the transformational, and the explanatory dimension and I argue that a theory produces knowledge through the epistemic cooperation of these dimensions. On the historiographical side, this approach entails that we have to pay more attention to the mathematical practices and to the way in which they shape the physical representation of phenomena.

Keywords Scientific theories · Received view · Semantic view · Models · Explanation · History and philosophy of science · Mathematical practices · Theory morphology

1.1 Rising Together or Sinking Together

History of science was born and raised as history of theories. The Copernican revolution, the mechanics of Galilei and Newton, the chemistry of Lavoisier, Einstein's special and general relativity, the quantum mechanics of Bohr, Heisenberg, and Schrödinger: these were the favorite themes of the scholars who pioneered the field. Particularly during the 1950s and 1960s—though important precursors can be found as early as the end of the nineteenth century—the historical research on scientific theories blossomed into a full-fledged, highly respectable academic discipline, which produced some of the finest pieces of cultural works of the last century and changed profoundly our picture of science. The boom of history of science benefited from

M. Badino, *The Bumpy Road*, SpringerBriefs in History of Science and Technology, DOI 10.1007/978-3-319-20031-6_1

a close, albeit often tacit, alliance with philosophy of science. The lively debate between the upholders of Logical Empiricism and its critics (which I survey below) provided a powerful arsenal of philosophical models, and above all guiding questions within which to understand the structure and the development of a scientific theory, particularly in physics and chemistry.

Unfortunately, the idyll was not to last. As historians accumulated study upon study, case upon case, archival research upon archival research, the connection with the philosophical models of theories began to fall apart. Personal letters, research notebooks, and first-hand reports all contributed to undermining the image of scientific theory as the highest peak of human logic and rationality. The break was mutually painful. Historians felt that philosophical accounts were no longer a reliable guide to understanding their data. Slowly but steadily, they diverted their attention away from theories and to other aspects of scientific research: laboratory life, academic interactions, institutional structures, pedagogical traditions, power relations, and hands-on practices. Those who remained interested in theories, replaced the philosophical vocabulary of models, axioms, observational terms, confirmation, corroboration, and so on with a language borrowed from anthropological approaches and cultural theory. Present historians of theory, thus, preferably talk about social context, nontextual dissemination, skills, and enculturation with the explicit aim to challenge the "prejudice that […] theories are isolated, ready-made products, which transcend the historical time and place of their making" (Galison and Warwick 1998, p. 288).

From their perspective, philosophers were divided about the attitude to adopt toward history of science. Many thought that, philosophy of science being the philosophical analysis of *contemporary science*, there was no real room for history. In addition, so the argument continued, the normative goals of philosophy of science simply do not match the descriptive aims of history, and thus it is no surprise that they fail to be of mutual help. Others, however, viewed the parting of history and philosophy of science as a dangerous step toward turning philosophical analysis into a self-authenticating—or, even worse, self-referring—intellectual construction totally detached from past and present scientific practices.[1]

Thus, if historiography and philosophy of theories have today parted ways, one of the reason is the failure of classical philosophical accounts of scientific theories. But yes, for better or for worse theories are an important part of everyday physical research, and yes, theories are intellectual constructions, precisely the kind of objects philosophy should help us deal with. Therefore, I believe that one of the very first bullet points in present philosophical agenda should be the development of a new framework able to capture the epistemic aspects of theories along with their historical evolution. In this chapter, I outline a proposal for such a framework. I begin with reviewing the philosophical debate on theories, from which I single out some deeply-ingrained assumptions. These assumptions, I argue, have prevented a more fruitful interaction between history and philosophy of science. By elaborating upon them,

---

[1] Giere (1973) set in motion a debate whose momentum never subsided; see, for instance, (Arabatzis and Schickore 2012; Burian 1977; Krüger 1982).

I propose to conceptualize theories as multidimensional structures concerned with three sorts of activities: representing, transforming, and explaining. In the ensuing chapters, I apply this conception to the analysis of Planck's radiation theory.

1.2 The Received View

True theories are all alike; every false theory is false in its own way. This slogan can be taken to summarize the gist of the so-called Received View of theories, i.e., the conception developed within the framework of Logical Empiricism. It would be historically inaccurate to consider this view as a monolithic doctrine, as it was elaborated by many actors over a long span of time.[2] However, it is fair to pinpoint some general theses common to Rudolf Carnap, Carl Gustav Hempel, Ernest Nagel, and other major contributors to the Received View.

The first part of the above slogan, 'true theories are all alike', leads immediately to two main points of this doctrine. First, theories ought to be analyzed by the means of logic. Famously, Logical Empiricism maintained that philosophy of science is concerned with the logical analysis of the propositions making up a scientific theory. In particular, Logical Empiricism draws a distinction between scientific terms used by scientists in their daily work (e.g., 'force', 'field', or 'evolution') and 'metascientific' terms used to talk about science such as 'law', 'explanation', and 'theory'. A philosopher should be interested in investigating the logic and the meaning of these latter terms, leaving the former to the scientist. As stated by Dudley Shapere, "to understand science, in the sense in which the philosopher of science wishes to understand it, would be to understand what is involved in saying that a certain scientific proposition or argument is a law, theory, explanation, etc." (Shapere 1984, p. 183). Now, of course, true laws, theories, and explanations are all alike and mutually replaceable as far as logic is concerned.

But there is a second point I have already mentioned above. For the Received View, theories are made up of true sentences that are axiomatically arranged. This means that a—relatively small—number of sentences is selected to be the axioms of the theory, and the remaining are logically derived as theorems. The vocabulary of the sentences is thus composed of logical and non-logical terms. The latter are divided into two categories: observational terms, which are directly connectable with immediate observation, and theoretical terms, which can be related with experience only through additional sentences called "correspondence rules" or "bridge principles". The correspondence rules provide the semantic of the theoretical terms. Hence, true theories are all alike because they are logically equivalent axiomatic systems.

The second part of my opening slogan synthesizes another crucial claim of the Received View. While there is only one way for a theory to be true, i.e., to correspond to a state of affairs in the world, there are many ways in which it can be false. To assemble a theory, the human mind makes a variety of choices, applies heuristic strategies, deploys tricks, or simply has sudden flashes of genius. None of these steps

---

[2]For a general introduction to the Received View see (Giere 2001; Mormann 2007; Suppe 1977).

can be systematized by an algorithm or a set of packaged rules. Because discovery is inherently a creative process, when it goes astray, it goes astray for different, unpredictable reasons. But Logical Empiricism was following a modernized version of the Cartesian dream of making philosophy a discipline of sound knowledge. To do this, one has to rely on a safe basis, i.e., immediate experience as the source of knowledge and logic as the tool to transform it. This double requisite marks the demarcational character of the philosophical program of Logical Empiricism: the prime task is to separate out genuine science from pseudo-knowledge.[3] Instead, discovery begins with analogies, models, heuristics, and preconceptions and proceeds by exotic means that have little to do with formal logic. For this reason, Hans Reichenbach distinguished neatly between the context of justification, which is the ambit of the logic of science, and the context of discovery (Reichenbach 1938). Ideally, for Logical Empiricism, the process of discovery should be the subject matter of cognitive science and psychology.[4]

The distinction between context of justification and context of discovery contains another important dualism introduced by Logical Empiricism. Any course of action performed by scientists, such as accepting, rejecting, or replacing a theory, is justified either by theoretical or by extra-theoretical reasons. In the first case, one accepts or rejects a theory, for instance, because of empirical evidence or because of logical connections with previously accepted laws, in brief, because of *internal* considerations. In the second case, though, the reasons to perform a theoretical choice are attributed to *external* factors such as personal preferences, career opportunities, political or social pressure and the like. Logical Empiricism links this distinction between internal and external factors to the concept of rationality. Thus, any course of action performed on the grounds of internal considerations is *eo ipso* rational: it is always rational to accept or reject a theory because of empirical observations or consistency problems. By contrast, it is irrational—and more often than not mistaken—to accept or reject a theory on the basis of political, social, ideological, or merely personal convenience. Hence, there is another sense in which theories can be falsified in multiple ways. For, if a scientist makes her moves for internal reasons only, she will automatically behave rationally and will turn sound knowledge into sound knowledge, whereas the adoption of one or many external factors of various natures will almost inevitably lead the theory astray. As a consequence, because truth is a sufficient reason to *believe* a true theory, reasons to believe a false theory must necessarily come from outside science. No rational agent would behave differently. This asymmetry of belief is characteristic of the internal/external distinction and leads to the 'sociology of error', i.e., the thesis that external factors might be invoked only to explain the belief in erroneous theories.[5]

---

[3]This strong emphasis on the demarcation between forms of knowledge is partly related to the specific historical and cultural situation of interwar Europe. See for instance (Stadler 2007) and the essays in (Giere and Richardson 1996).

[4]Recently, the analysis of discovery has become increasingly popular also for its relations with Artificial Intelligence (Meheus and Nickles 2009; Nickles 1978, 2001).

[5]On this point see for instance (Barnes 1974).

From this cursory survey, it appears that the gist of the program of Logical Empiricism is the clarification of what a theory boils down to by focusing upon its *logical form*. All meaningful philosophical issues concerning theories can be settled by the analysis of how sentences stay together and support, contradict, or refute each other. Conversely, if an issue cannot be settled in this manner, it is meaningless.

Because of its intense focus upon the logical form, the Received View has been sometimes referred to as the 'syntactic approach to theories'. In fact, semantics play here only a subordinate role. Generally, the meaning of the theoretical terms is established unambiguously and directly by means of the relation with observational terms defined by the correspondence rules (Carnap 1956). This simple semantic, though, generates scores of problems. To begin with, it presupposes a neat distinction between observational and theoretical terms, a claim that has been severely criticized on the grounds that there is no theory-free access to immediate experience. In addition, this semantic seems to entail that, after all, we do not really need theoretical terms. Ideally, observational terms give complete and explicit definitions of theoretical terms, and therefore, the latter are, in the best case, simply a shorthand formulation of the former. This consequence was logically proven and is known as Craig's theorem (Craig 1953). To be sure, even before Craig, Carnap had already envisaged the reductionism encapsulated in the notion of explicit observational definition of theoretical terms and introduced the alternative doctrine of *partial interpretation* (Carnap 1936–1937). According to Carnap, observational terms do not exhaust the semantic of theoretical terms, but rather "correspondence rules only *partially* define them because more than one reduction sentence (correspondence rule) is possible for the same theoretical term" (Suppe 1977, p. 22). However, this strategy leads to what Hempel has dubbed the 'theoretician's dilemma':

> If the terms and principles of a theory serve their purpose, they are unnecessary [...] and if they do not serve their purpose, they are sure unnecessary. But given any theory, its terms and principles either serve their purpose or they do not. Hence the terms and principles of any theory are unnecessary. (Hempel 1958, pp. 49–50)

Furthermore, it was soon discovered that Carnap's notion of partial interpretation was not as straightforward as it appeared at first sight. Peter Achinstein and Hilary Putnam proposed a list of possible readings of this notion and declared all of them to be untenable (Achinstein 1963; Putnam 1962).

Additionally, the strong demarcational ambition of the program introduced several problems. The Received View aims at capturing all relevant elements that make an array of sentences a genuine scientific theory through axiomatization. What does not conform with this requisite is not a theory. However, unless one cannot exhibit a priori or transcendental arguments for this claim, it is necessary to have some preconceived notion of theory that allows us to establish, for instance, that we should be worried about not capturing quantum mechanics, but not so worried about neglecting Freudian psychoanalysis. Thus, there seems to be a sort of circularity in the ambition of the Received View to grasp the essence of a scientific theory.

The difficulties of the Received View, and this latter point in particular, reveal a more general tension that crosses all philosophical accounts of scientific theories. In the first half of the twentieth century, many writers saw in the formal methods the only way to escape a speculative approach to philosophical problems, what they scornfully labeled as metaphysics. The hope was cherished that the use of mathematical logic could put a halt to unfruitful and inconclusive debates by unearthing the real structure of scientific theories. For Logical Empiricism, a theory is a totally transparent object in which everything that truly counts is visible and formalizable in an axiomatic system. The logical and non-logical vocabulary, the relations between terms, and the way to combine them are fixed once and for all. But, as Da Costa and French have brilliantly put it, "the dangers of such an approach are well known: seduced by the scholastic angels dancing on the formal pinhead, we lose sight of the practice we are trying to understand."[6] Formal methods can give us only the *illusion* to solve all problems, by turning theories into philosophical artifacts that are worlds away from the actual scientific theories we all know and love. In the 1960s, history-oriented philosophers of science such as Thomas Kuhn and Paul Feyerabend argued that, far from being transparent, scientific theories are in fact very opaque objects relying on tacit knowledge, pedagogical drills, inculcated methods, shrewd rhetoric, and—at times—power relations. Dudley Shapere has called 'global presuppositionism' the thesis that the best part of scientific knowledge is not stored in its logical form, but rather in unformalizable practices (Shapere 1984, p. XVI).

As we will see in the next sections, here lies one of the fundamental dualisms marking the philosophical efforts to explicate scientific theories. On the one hand, there is the call for rigorous arguments, a demand best fulfilled by the deployment of formal methods; on the other, there is the equally compelling necessity to capture actual scientific theories instead of philosophical straw men. After the failure of the Received View, philosophers tried to progressively relax some of the tight constraints of mathematical logic and to better approximate concrete scientific practices.

1.3 The Semantic View: Models and Their Discontents

Although the axiomatization of theories proposed by the Received View was powerful, to many it appeared as excessively artificial. It is very unusual, even in mature sciences, to be able to isolate a set of axiomatic sentences. Simply, scientists do not proceed that way. Rather, they think in terms of systems that concretely realize their abstract theories. Following this thought, at the end of the 1950s, many philosophers suggested that the emphasis of analysis should be shifted from syntax to semantics and from sentences to models.[7] After borrowing my first slogan from Tolstoy, I will

---

[6]Da Costa and French (1998), p. 125. One of the first to point out that theories are generally much richer objects than assumed by the Received View was (Achinstein 1968).

[7]A pioneer of this approach was Evert Beth, who proposed a semantic for theories at the end of the 1940s (Beth 1948, 1960; Van Fraassen 1970). For a general discussion of the Semantic View see (Giere 2001; Suppe 1989; Van Fraassen 1987).

now formulate a second by adapting Jane Austen: it is a truth universally acknowledged, that a single scientist in possession of a good theory must be in want of a model.

The concept of model is famous for its elusiveness.[8] In his early contributions to the Semantic View, Patrick Suppes insisted on the necessity of adopting the Tarskian notion of model as a realization of the theory (i.e., as a set-theoretical entity that makes the sentences of the theory true) and pointed out the ensuing novelty of the Semantic View: "the important distinction that we shall need is that a theory is a linguistic entity consisting of a set of sentences and models are non-linguistic entities in which the theory is satisfied" (Suppes 1969a, p. 13).[9] According to Suppes and others, scientists do not work with axioms, but rather with concrete systems that embody the theory. Thus, although mechanics can be formulated as a self-contained set of sentences connected by logical relations, in practice it is worked out as a theory of *mechanical systems*, i.e., colliding particles, oscillating pendulums, orbiting planets and so forth. We surely get closer to how scientists really work if we formulate our philosophical analysis in terms of models of theories and models of data (i.e., set-theoretical structures representing the results of experiments): "the attempt to characterize exactly models of an empirical theory almost inevitably yields a more precise and clearer understanding of the exact character of the theory" (Suppes 1969a, p. 18).

In addition, and this point has not been adequately stressed by commentators, models open up a more historical perspective on the study of theories. The very term 'model' points at different stages in the development of a discipline:

> In old and established branches of physics which correspond well with the empirical phenomena they attempt to explain, there is only a slight tendency ever to use the word 'model'. The language of the theory, experiment and common sense is blended into one realistic whole. Sentences of the theory are asserted as if they are the one way of describing the universe. Experimental results are described as if there were but one obvious language for describing them. [...] On the other hand, in those branches of physics which give as yet an inadequate account of the detailed physical phenomena with which they are concerned there is a much more frequent use of the word 'model'. (Suppes 1969a, p. 15)

Thus, from the early 1960s, philosophers of science began to argue that a deeper attention to theories' semantic components would bring philosophy closer to the actual practices of scientists. At the same time, relying on Tarski's account endowed the analysis with the powerful formal methods of set theory. This program was effectively synthesized by Suppes' slogan "to axiomatize is to define a set-theoretical predicate" (Suppes 1969b). Set theory replaced logic of predicates, and the key concept of isomorphism replaced logical consequence. The highest peak of formalization in the framework of the Semantic View was the *representation theorem*: "to establish a representation theorem for a theory is to prove that there is a class of models of the theory such that every model of the theory is isomorphic to some member of this class" (Suppes 1969a, p. 17).

---

[8]For a recent discussion see (Bailer-Jones 2009).
[9]See also (Suppes 1967).

The Semantic View allows for more flexibility in the analysis of theory structure, while at the same time maintaining the usage of the rigorous methods of set theory. It is fair to say that it is still the most popular view of theories among philosophers who retain a formal attitude to the issue. However, the approach still falls short of its original goal, of bridging the gap between philosophical discussion and scientists' actual practice. For instance, if it is true that scientists seldom adopt propositional axioms, it is as true that they do not regularly deploy Tarskian models. More often than not, models find a place in scientific practices because of their visual, heuristic, or suggestive features. Billiard balls do not represent *exactly* gas molecules and do not behave *exactly* like them; in other words the billiard ball model does not make the sentences of kinetic theory literally true. The reason why it remains very popular lies in what Peter Achinstein called the *iconic* value of models (Achinstein 1968): the capability to illustrate the basic idea with one evocative image. Therefore, it seems that the insistence on a set-theoretical notion of model chiefly leads to the replacement of one arsenal of formal methods with another, rather than to the production of a more plausible account of theories.

This consideration goes hand in hand with another point: if models are merely alternative interpretations of a theory, why do we need them in the first place? Why do we have to start with an interpretation rather than the real thing, i.e., the theory as it is historically given on paper? Is it not true that, like in the case of Logical Empiricism, the imperative of formalization creates yet another philosophical artifact? This problem becomes even more urgent when one considers that models often encourage a confusion between their domain of application and that of the original theory (Da Costa and French 1998).

However, the strongest critical point raised against the Semantic View concerns the alleged relation between models and theory. Within the framework of the Semantic View, a theory may have many models, but a model is such only in relation to a theory: there is no such thing as an interpretation detached from what is interpreted. This asymmetry makes models parasitic on theories, which is the reason for the previous point about the dispensability of models. However, in the late 1980s and especially in the 1990s, many philosophers of science insisted that models are much richer, complex, and autonomous entities than the Semantic View is willing to acknowledge.

These objections and the emergence of a more historically-minded philosophy of science championed by Kuhn and Feyerabend brought about dramatically new developments in the philosophical accounts of theories. I discuss these positions in the next section.

1.4 Development of the Semantic View

Although the Semantic View remains very popular, philosophers of science have grown increasingly skeptical that the set-theoretical formalism is the adequate—let alone the only—way to deal with models. Two main threads of research have emerged out of this release of the formalistic grip. Many scholars have tried to work on less formal aspects of models such as their being (1) specific and problem-oriented, (2) representational devices, and (3) mediators between theory and data. This work has required much discussion of the actual practices of scientists. Other writers have shifted their focus from the analysis of theories to their evolution over time and their being part of traditions or research programs. This line, of course, is closely intertwined with historical research.

1.4.1 Many Faces of Models

I begin my discussion with what I call the sophisticated version of the Semantic View, which is a cluster of approaches developed since the late 1970s by philosophers such as Nancy Cartwright, Ronald Giere, Mary Morgan, and Margaret Morrison among others. These authors are interested in developing an account of theories closer to actual scientific practices and overcoming a certain tendency to abstractness typical of previous approaches.

1.4.1.1 Giere

A common trait of the work of Cartwright and Giere is the criticism of a deep-seated assumption of philosophy of science: the claim that the analysis of theories should be concerned only with fundamental laws, while the approximations needed to put these laws to work on concrete systems have a mere auxiliary role:

> Those scientists and philosophers who have taken it for granted that the laws of nature are well confirmed, general statements have obviously not been ignorant of the fact that scientists regularly use approximations. But they have taken this to be a relatively inconsequential fact about science. They have regarded the fact as a matter of only practical, not theoretical importance. (Giere 1988, p. 78)

Giere's argument to challenge this view relies on the examination of the scientific practices coded in textbooks, the main tools through which scientists become acquainted with their discipline. He especially focuses upon the most venerable among the physical theories, i.e., classical mechanics. Giere pays scarce attention to the worries Kuhn and others have voiced against the suspiciously polished picture of science conveyed by textbooks: his task is not to understand history, but rather to analyze how contemporary science is learned and practiced. Giere's discussion of some popular textbooks comes to the conclusion that, *contra* Logical Empiricism, the fundamental laws of mechanics are never considered as axioms whose logical consequences have to be connected with observations through correspondence rules. Although this is a consistent reconstruction of what a theory looks like, it does not seem to take place in the learning process of scientists. In fact, taken as empirical statements, the laws of motion must be regarded as false: "the most general 'laws' of mechanics, like $F = ma$, are not really empirical claims, but more like general schemas that need to be filled in with a specific force function" (Giere 1988, p. 76).

Thus, after introducing the laws of motion, physics textbooks present a battery of exemplary cases in which the force function takes particularly interesting forms: the falling body, the linear oscillator, the pendulum, and so on. These cases remain simple due to a set of simplifying assumptions: the spring of the oscillator swings without friction, the pendulum makes only small oscillations, and so on. The cases can be complicated further by introducing frictions, non-negligible masses or angles, tensions, and other factors. These complications enrich the model and generate a problem of *interpretation*, that is, the connection between mathematical symbols and the physical quantities of the model. Quite a different problem is to link the model with specific, physical objects. For instance, one thing is to interpret a symbol as the velocity of a planet in a three-body model, another thing is to identify that velocity as the velocity of the Moon. To do that, one needs a *theoretical hypothesis*, "a *linguistic* entity, namely a statement asserting some sort of relationship between a model and a designed real system" (Giere 1988, p. 80).

Thus, while the philosopher's theory is a set of statements interrelated by logical rules of inference and connected to observational statements by correspondence rules, the scientist's theory is a set of models—nonlinguistic entities that make true some set of equations—interrelated by some relations of *similarity* and in turn connected to real systems by means of other relations of similarity substantiated by suitable approximations. Although Giere does not commit himself to a precise definition of theory, he concludes that "we understand a theory as comprising two elements: (1) a population of models, and (2) various hypotheses linking those models with systems in the real world" (Giere 1988, p. 85). I want to stress two points about this account.

First, Giere's prime aim in embracing the Semantic View is not to use the weaponry of set-theoretical methods. He is not interested in the characteristics of the logical structure of models. He does not explicitly deny that a clarification of these features can be helpful, but he does not restrict his discussion to a special relation between models. He criticizes Van Fraassen's suggestion that such a relation should be isomorphism, because isomorphism does not cover the variety of relations that scientists deploy (Van Fraassen 1980). For Giere, the whole point of adopting models instead of axioms and statements is that models have a *representational* power. This is what a theory is supposed to do: it must represent the world by specifying the force function in artificial constructions that can be made more and more complicated as well as more and more similar to the external physical world.

Second, following the set-theoretical tradition, Giere is willing to express the relation between laws and models in terms of truth, but with an important addendum:

> The relationship between some (suitably interpreted) equations and their corresponding model may be described as one of characterization, or even definition. We may even appropriately speak here of 'truth'. The interpreted equations are *true of* the corresponding model. But truth here has no *epistemological* significance. The equations truly describe the model because the model is defined as something that exactly satisfies the equations. (Giere 1988, p. 79)

Here, the key point is the epistemological significance of truth. Giere seems to mean that our knowledge does not advance much by being informed of the fact that some laws are true for a certain model. But to fully appreciate the value of this point, the best place to look is Nancy Cartwright's work.

1.4.1.2 Cartwight

To increase our knowledge of the external world, we do not only need to accumulate information about facts of nature, but we also need to connect them properly. Briefly said, we have to explain facts and behaviors by means of other facts and behaviors. And, it almost goes without saying, an explanation must be based on *true* facts and behaviors. No false explanation is worth the name. However, not every true sentence can be used to explain. We can say a bunch of true things about Paul Atreides, but none of them can feature in any explanation of natural phenomena, because they are true only in Frank Herbert's *Dune*. This is what, I think, Giere means in the quote above: a true sentence is epistemologically significant when it can be part of an explanatory argument and therefore increase our knowledge of the world. Cartwright puts the point even more forcefully when she provocatively argues that "truth does not explain much".[10] One should distinguish between two kinds of laws. On the one hand, there are the so-called phenomenological laws, complicated characterizations of specific facts obtained by introducing several constants, parameters, and approximations. Those laws are truthful descriptions of the facts they target. On the other hand, there are fundamental laws such as Newton's law of gravitation, the second principle of thermodynamics, or even Boltzmann's transport equation. These are beautifully simple and general laws able to cover many facts, but none of them is literally truth. These laws, which actually feature in explanations, are true *ceteris paribus*, that is in ideal conditions that leave out a large part of the messy world:

> Most scientific explanations use *ceteris paribus* laws. These laws, read literally as descriptive statements, are false, not only false but deemed false even in the context of use. This is no surprise: we want laws that unify; but what happens may well be varied and diverse. We are lucky that we can organize phenomena at all. There is no reason to think that the principles that best organize will be true, not that the principles that are true will organize much. (Cartwright 1983, pp. 52–53)

This point brings us straight to Cartwright's discussion of explanation. According to the deductive-nomological model (D-N model) developed by Hempel, fundamental laws tell us how the world really is; hence, they are true. By contrast, phenomenological laws are approximated and therefore far from truth. The former explain the latter by the addition of suitable initial conditions. Cartwright turns this thesis upside down. First of all, approximations do not take us further from truth, but closer! Fundamental laws are false of the real world, therefore we can come to know something true only by relaxing the *ceteris paribus* clause, which means to introduce complicated approximations. Like Giere, Cartwright also highlights the

---

[10]Obviously, this thesis flies in the face of the inference to the best explanation, see (Cartwright 1983, pp. 4–10).

epistemic role of approximations conspicuously denied by Logical Empiricism.[11] Secondly, she distinguishes two components of scientific explanation. To explain means, partly, to provide a causal account of a phenomenon. For obvious reasons, the sentences stating the causes must be true; therefore, this is a task for the highly specific and detailed phenomenological laws. But the second aspect of the explanatory activity concerns the embedding of the phenomenon in a theoretical net, i.e., to establish relations with other phenomena. This task cannot be accomplished by phenomenological laws, and it is therefore the prime goal of fundamental laws.

From the perspective of this distinction, the D-N model is insufficient. Cartwright proposes to replace it with the 'simulacrum account' of explanation (Cartwright 1983, pp. 143–162). The first stage of the explanatory activity is an unprepared, very rich description of the phenomena, with a lot of details. From this unprepared description, one distillates a prepared one, with only some relevant elements. This considerably simpler entity is a model. The model is conceived by taking into account the fundamental laws and their *ceteris paribus* range of application. To see a rich phenomenon as a species of a more general model allows for the required unification with other phenomena. The consequence, however, is that the fundamental laws, which are applicable to the model, *are true of the objects of the model, not of the objects of nature*. This is how explanation is reached at the expense of truth. To the extent that fundamental laws explain, they are not true (of nature), and to the extent that phenomenological laws are true, they do not explain (in the unification sense).

Two larger points emerge from Cartwright's position. First, Cartwright wants to capture the basic intuition that the world out there is messy and complicated and cannot be harnessed by fundamental, simple laws. An adequate understanding of the world requires approximations, qualitative approaches, heuristics, and good sense. Our theories are made up of useful models, but we must be aware of the epistemological distance between those models and the real world. Too formal an attitude tends to cancel out that distance and to regard models as more explanatory than they actually are. Second, Cartwright, more emphatically than Giere, puts her finger on the important relation between an account of theories and explanation. We do not construe theories for the sake of it: we do it because we want to explain the world. Logical Empiricism was all too eager to treat these questions separately, but this strategy also reveals its leaning toward abstraction: as a matter of fact, the practice of theory construction is inextricably intertwined with the practice of explaining.

1.4.1.3 Models as Mediators

Before discussing this practice, though, there is another development of the Semantic View that is worth considering. One assumption taken for granted by the upholders of the Semantic View is that models do not make much sense outside a theory:

---

[11] As a side remark, it is interesting to note that Cartwright's use of textbook science is somewhat opposite to Giere's. She also stresses that a theory is presented in textbooks as a sequence of models, but then she concludes that such a presentation is detached from the real world.

models are always at the service of a theory. At the end of the 1990s, historians and philosophers of science started to increasingly dispute this assumption. In an edited volume published in 1999, Margaret Morrison and Mary Morgan collected several studies supporting the claim that models are autonomous entities:

> [M]odels, by virtue of their construction, embody an element of independence from both theory and data (or phenomena): it is because they are made up of a *mixture* of elements, including those from outside the original domain of investigation, that they maintain this partially independent status. (Morrison and Morgan 1999, p. 14)

They go on arguing that models are independent of theories in two crucial respects. On the one hand, they are functionally autonomous because they can be developed even when no theory is available. The case of the London-London model of super-conductivity or Prandtl's model of fluid (one of Morrison's favorite examples) show that models can work well without the support of any theory. On the other hand, they are also representationally autonomous because they are able to unfold relations and dependencies between phenomena, which make them more suitable for representing the external world. The key feature for successfully performance of this array of functions is the mediating character of models. They connect disparate entities from different sources, and in so doing, they tell us something about each of these sources. The point by Morrison and Morgan that "models function not just as a means of intervention, but also as a means of representation" (Morrison and Morgan 1999, p. 12) is extremely valuable and needs to be spelled out and generalized for later use. Their argument is that there is a sense of instrumentality that is closely related to representation. Basically, some instruments used "as investigative devices" teach us something about the object they intervene upon. I submit that this is not an exclusive feature of models: instead, any device that allows for a symbolic manipulation provides a certain amount of representational knowledge; it has, in other words, an epistemic content. This thesis will come in handy very soon.

1.4.2 Theories Change

We have seen that, starting from the late 1960s, philosophers of science progressively relaxed the strict formal requirements of the Received View. This evolution was due to two concomitant factors. First was the necessity of bringing the philosophical accounts closer to the actual scientific practices. Second—and to be sure this factor was active long before the first one—was the realization that the classical approaches were hopelessly in contrast with the data that historians of science had been accumulating during the 1950s and 1960s. The formidable historical effort produced in the heydays of the discipline painted a picture of scientific theories very far from the orderly axiomatic system dreamt of by the upholders of the Received View. Galilei, Newton, Faraday, and Einstein did not appear to start with a set of fundamental statements, to distinguish neatly between theoretical and observational terms, or to be willing to give up a theory in the face of the slightest empirical difficulty. Instead, they seemed to stubbornly stick to their beliefs, to pay attention to their social relations with the rest of the scientific community, and to blend their scientific concepts with a plethora of broader cultural elements.

While some philosophers, notably Kuhn and Feyerabend, came to the conclusion that science was inevitably beset with (or blessed by) irrationality to various degrees, others felt that their own task was to get their hands dirty in the turbulent magma of history and to separate the wheat of the purely rational changes from the chaff of irrationality. Philosophical accounts should reflect the fact that science has a "historical character" (Lakatos 1978, p. 35) while at the same time keeping the requisite of science as a rational endeavor.

1.4.2.1 Lakatos

Imre Lakatos was among the first to take up this challenge. He realized that the historical fact that theories sometimes survive apparently (and retrospectively) decisive empirical confutation is philosophically mirrored by the conventionalist thesis that one can always save a theory by tinkering with definitions and auxiliary hypotheses. This claim still leaves open the question of distinguishing between genuine, rational rescues and theoretically possible, but ultimately barren, semantic adjustments. Thus, Lakatos' central question: "[w]hy not rather impose certain standards on the theoretical adjustments by which one is allowed to save a theory?" (Lakatos 1978, p. 32).

Lakatos introduces a major shift in philosophy of science: he moves the focus from individual theories to series of theories or, in his jargon, research programs. A research program is made up of two ingredients. First, there is the hard core of the program, a set of assumptions, statements, laws, and hypotheses that cannot be touched without undermining the conceptual integrity of the program itself. Lakatos' way to guarantee that a series of theories belong to the same research program is to require that they share a common core of statements that characterizes the *principium individuationis* of the program. He calls 'negative heuristic' the set of strategies aiming at keeping the hard core safe from any attempt of confutation.

The second ingredient allows for a temporal change in the theories making up the research program, that is the protective belt: a series of auxiliary hypotheses that can be adjusted, added, or abandoned in order to bring the core in accord with observations. The 'positive heuristic' is concerned with the possible modifications of the protective belt. It is important to realize that, although Lakatos' account concerns primarily research programs, these programs and theories do not differ dramatically. Take, for instance, the Newtonian research program: the essence of the program is the essence of any good Newtonian theory:

> In Newton's program the negative heuristic bids us to divert the *modus tollens* from Newton's three laws of dynamics and his law of gravitation. This core is 'irrefutable' by the methodological decision of its proponents: anomalies must lead to changes only in the 'protective' belt of auxiliary, 'observational' hypotheses and initial conditions. (Lakatos 1978, p. 48)

Present-day versions of Newtonian mechanics, which surely differ greatly from the original one, still can claim the same hard core, although with a very different protective belt. It follows that individual theories display the same morphology of 'theoretical core and protective belt'. While this simple morphology warrants the continuity of the research program, its rationality is warranted by the distinction between progressive and regressive problem shifts. Any change in the protective belt can lead to an increase or to a decrease of knowledge. Obviously, it is rational to push forward a theory only when such a move promises to increase our knowledge.

1.4.2.2 Laudan

Lakatos' crucial question and his approach to tackling it was very influential on philosophy of science. Larry Laudan, for one, extended Lakatos' idea of research program to more comprehensive research traditions. Laudan's strategy is to distinguish precisely between the ontological and methodological assumptions implicit in a research tradition and the concrete realizations of these elements in individual theories. A research tradition is not characterized by scientific laws or principles, but by assumptions about what the objects are and how we should investigate them:

> In brief, a research tradition provides a set of guidelines for the development of specific theories. Part of those guidelines constitute an ontology which specifies, in a general way, the types of fundamental entities which exist in the domain or domains within which the research tradition is embedded. The function of specific theories within the research tradition is to explain all the empirical problems in the domain by "reducing" them to the ontology of the research tradition. (Laudan 1977, p. 79)

As for the conception of theories, Laudan's stance is more flexible than the Semantic View. He sees one of the characterizing traits of theories in their being answers to problems. This move allows Laudan to shift the analysis from theories to the problems they are supposed to solve. Thus, for instance, he distinguishes between conceptual and empirical problems as two different sources of difficulty. The other characterizing trait of theories is that they essentially are representational devices: theories are organized systems of statements and models able to represent the world as we experience it. The combination of these two elements—by no means exclusive of Laudan's stance—reveals the wide-spread *intellectualistic bias* of the philosophical accounts of theories. The underlying thesis that theories solve problems by suitably representing the world is challenged by the basic considerations that representations do not solve problems, *practices do*. According to the intellectualistic bias, the epistemological arrow goes from an adequate representation straight to a solution implemented by epistemically neutral practices. No sooner than the problem is represented, it is solved. Another way to express the intellectualistic bias is the claim that there is an *epistemic asymmetry* between the various elements of a theory, with the representational elements at the top of the pyramid. I dispute this thesis, and in what follows, I argue that the epistemic content of a theory is shared among its different constituents.

As far as the morphology of theories is concerned, Laudan does not go remarkably further than Lakatos' core-belt schema. He also believes that "certain elements of a research tradition are more central to, more entrenched within, the research tradition than other elements" (Laudan 1977, p. 99). These elements form the 'unrejectable core' of the research tradition, the elements without which a theory is outside the tradition itself. Contrary to Lakatos, however, this core can occasionally change over time. Its elements are not obliterated, but they move from central to peripheral positions:

> During the evolution of any active research tradition, scientists learn more about the conceptual dependence and autonomy of its various elements; when it can be shown that certain elements, previously regarded as essential to the whole enterprise, can be jettisoned without compromising the problem-solving success of the tradition itself, these elements cease to be a part of the "unrejectable core". (Laudan 1977, p. 100)

This process opens a Pandora's box of how to conceptualize scientific revolutions, a topic about which Laudan only offers incomplete hunches.

1.4.2.3 Shapere

As we have seen, many upholders of the core-belt morphology also subscribe to what Dudley Shapere has dubbed 'the inviolability thesis': "the idea that there is something about the scientific (or, more generally, the knowledge-seeking or knowledge-acquiring) enterprise that cannot be rejected or altered in the light of any other beliefs at which we might arrive, but that, on the contrary, must be accepted before we can arrive, or perhaps even seek, such other beliefs" (Shapere 1984, pp. xix–xx); see also (Shapere 1987). Note, however, that this connection is more an effect of the strong cultural influence of Logical Empiricism than a logical implication. For instance, Shapere's 'bootstrap conceptualism' combines the core-belt morphology with the pragmatist axiom that any belief can be doubted, but not all beliefs can be doubted at the same time. Shapere argues that a distinction between untouchable and revisable knowledge remains indispensable for science, but he challenges the claim that the honorific title of 'essence' should be bestowed upon this knowledge. Instead, we should form our background knowledge on the basis of our best available beliefs in current science and, accordingly, it can be thrown away when superseded by better beliefs. Shapere does not see the need of a *principium individuationis*, a fixed semantic and conceptual identity of a theory/research program/research tradition for science to work properly. But apart from this—truly remarkable—difference, Shapere agrees with other writers that theories are essentially answers to problems. As we have seen, this amounts to a combination of a taxonomy of problems, with a conditional criterion: 'if one has the problem of kind P, then it is necessary to construct a theory of kind T', see (Shapere 1984, pp. 286–287).

1.5 The Whewell Thesis

The foregoing analysis has highlighted four general points: (1) in the course of time, philosophers have re-evaluated the epistemological role of practices such as approximations, even if (2) the prime focus is still upon concepts and laws; (3) there is an important relation between theories and explanations, as theories are (at least partially) explanatory devices; and (4) instruments and tools can have a representational value. However, it has also revealed three enduring assumptions of scientific theories: (A) there is a tendency to treat theories as a given, (B) philosophical accounts of theories rely on a Lakatosian core-belt morphology, and (C) the thesis of the epistemic asymmetry that exists between the constituents of a theory. My goal for the remainder of this chapter is to criticize the assumptions (A)–(C) and to elaborate on points (1)–(4) to build an alternative approach.

As already anticipated at the end of Sect. 1.2, if we look back at the development of the philosophical reflections of scientific theories, we can easily discern a general pattern of evolution. Both the Received View and the more technical versions of the Semantic View make large use of formal methods. Clearly, they want to guarantee a high standard of rigor in the argumentation, and their goal is to distinguish systematically between a scientific theory and a unstructured collection of utterances. Although these proposals meet their philosophical aims, they fail to represent adequately the complexity of scientific practices. The disappointing result is that the final products are closer to philosophical artifacts than to real theories.

Reactions to this abuse of formal methods began to crop up in the late 1960s. As we have seen, several scholars progressively relaxed formal requisites and tried to build accounts closer to scientists' actual practice. Philosophical views of theories become more historical and social, but at the same time they lose interest for the internal structure to the point of conflating theories with the problems they are supposed to answer. In both cases, theories are treated as a given. In the formal approaches, they are decomposed into some elementary constituents (propositions, models), but they are essentially a-historical objects whose origins are not philosophically relevant. In the historical-social approaches, theories acquire a temporal dimension, but they are deprived of their internal structure: they become the elementary constituents of research traditions. So, it seems that philosophers of science systematically make a choice between describing the internal structure of theories or their evolution over time. The conclusion has been synthetically summarized by Shapere: "There is today no completely—one is almost tempted to say remotely— satisfactory analysis of the notion of a scientific theory" (Shapere 1984, p. 112). This is what I mean by assumption (A) above: theories have been treated alternatively as a historical or a structural given. It is important to stress that this point is related to assumption (B). A moment's reflection reveals that even the formal approaches generally accept implicitly the core-belt morphology. Both the Received View and the Semantic View still distinguish clearly between central ingredients of the theory (axioms, exemplary models) and the peripheral elements related to the center by means of logical or semantical connections. The history-oriented approaches take up the same morphology and simplify its internal structure to adapt it to the explanation of temporal development.

Finally, by assumption (C) I mean the following, very simple claim. Scientific theories—at least physical theories, which are the subject matter of this book—are made up of many elements: concepts, assumptions, laws, principles, ideal objects, hypotheses, models, approximations, and mathematical techniques. Although a theory is such a complex and multifaceted entity, its knowledge content has been traditionally seen as stored in very few of its constituents. It is usually assumed that there is a theoretical core that makes a theory what it is and that contains everything epistemically relevant a theory has to say. In particular, mathematical techniques are not viewed as contributing to the epistemic content of a theory. They simply serve for specific applications.

Assumptions (A), (B), and (C) are another expression of the persistent intellectualistic bias in philosophy of science, an attitude I here call *the Whewell thesis*. This is not because William Whewell was the first to enunciate it, but simply because his formulation synthetically displays its implications. The Whewell thesis can be found in his Bridgewater Treatise dedicated to *Astronomy and General Physics*. Famously, the purpose of the book was to defend the idea of a divine designer against the physicalistic interpretation of celestial mechanics given by many French mathematicians, notably Laplace. Toward the end of the book, Whewell strategically contrasts the believer Newton with the atheistic Frenchmen and establishes a hierarchy on the basis of the epistemological value of their work. He argues that the fundamental laws of nature, such as the law of gravitation, "occupy little room in their statement", but that, once formulated, those laws contain all the knowledge possibly provided by the theory. The subsequent drawing of the consequences by means of mathematical techniques, albeit ingenious, does not add a single bit of knowledge:

> In explaining, as the great [French] mathematicians just mentioned have done, the phenomena of the solar system by means of the law of universal gravitation, the conclusions at which they arrived were really *included in the truth of the law*, whatever skill and sagacity it might require to develop and extricate them from the general principle. (Whewell 1837, p. 328), emphasis added

In closing the chapter, Whewell formulates his thesis even more outspokenly:

> When, therefore, we consider the mathematicians who are employed in successfully applying the mechanical philosophy, as men well deserving of honor from those who take an interest in the progress of science, we do rightly; but it is still to be recollected, that in doing this they are not carrying us to any higher point of view in the knowledge of nature than we had attained before: they are only unfolding the consequences, which were already virtually in our possession. (Whewell 1837, pp. 332–333)

The Whewell thesis proposes a view of theories according to which the mathematical practices simply unfold an epistemic content stored *ab initio* in a few fundamental laws. Ultimately, it imposes an unjustified epistemic asymmetry between some theoretical elements and the rest of the ingredients of a theory, particularly its mathematics. Mathematical tools can evolve, or it might happen that, in the course of time, techniques previously neglected suddenly acquire importance. Though, they only unravel what is "virtually in our possession" in the theoretical core. Using perturbation theory or topology adds nothing to the knowledge content of gravitational laws; it simply allows for a faster, more elegant, or more effective way to derive the consequences. Traditionally, this claim has been tacitly justified by the belief that mathematical practices do not strictly belong to a physical theory per se, but are imported from pure mathematics as an aid to express the physical content in a workable manner and to allow for wide unifications with different branches of the physical research.[12] However, this belief is historically unjustified: Newton's mechanics does not just consist of the three laws of dynamics and the law of gravitation, but also in all the techniques to put them to work, to turn them into equations of motion, to manipulate and—when possible—to solve those equations. Furthermore, it is also conceptually misguided, because it does not entail the epistemic neutrality of mathematical practices. Thirdly, and finally, the Whewell thesis interferes seriously with the elaboration of a philosophically and historically satisfactory account of the process of theory construction i.e., with the reconciliation between an analysis of the internal structure and its temporal development. By distinguishing crudely between a receptacle of the whole knowledge content and a large amount of epistemically neutral auxiliary tools, the Whewell thesis makes the generation of a theory a totally mysterious event. The theoretical core does not receive its content from anything; it pops up from a flash of genius or from exotic psychological operations generically labeled as 'context of discovery'. This lack of a plausible explanation for the generation of the theoretical core has made available plenty of room for sociological accounts, which, albeit interesting, are not the whole story.

1.6 An Historical-Epistemological Workspace

1.6.1 Beyond the Whewell Thesis

My strategy to replace the Whewell thesis is to extend point (1) and, at the same time, to deny point (2) above. We have seen that philosophers of science have progressively recognized that approximation techniques have an important epistemic role to play, although the main focus has remained upon a handful of fundamental laws. I propose that we should recognize also an epistemic relevance to other theoretical components. To ease the conceptualization of the Whewell thesis and its epistemological implications, I submit that the epistemic job of a theory can be divided into three dimensions, which concern what we expect from it.

Firstly, there is the *representational dimension*, which has to do with the ability of the theory to represent the external world by means of a selection and organization of the relevant aspects and their expression in symbolic terms. In other words, the theory yields a *symbolic codification* of physical phenomena. Laws, models, ideal objects,

---

[12]For a recent discussion of this point see (Morrison 2000).

concepts, and physical quantities all contribute to the representative dimension of the epistemic job. For instance, both the laws of dynamics and the model of linear oscillator prevalently aim at describing a class of phenomena. This does not exclude, however, that they might also be used for other purposes at other times or in other contexts.

Secondly, a theory possesses a *transformational dimension*, i.e., the function of manipulating and transforming symbolic codifications to obtain solutions to specific problems in terms of other symbolic codifications. Because codifications are mathematically expressed, usually their transformation is carried out by a package of mathematical tools and techniques or, more generally *symbolic practices*. The symbolic transformation can be performed in several ways, which may be mathematically equivalent, but often have different physical consequences.

Thirdly, we also expect a theory to have an *explanatory dimension*, in the sense that it must produce a certain kind of explanation of the physical behaviors in terms of some reasons and some mechanism (the why and how of the phenomena). This dimension opens up the thorny philosophical problem of the nature of explanation whose discussion I postpone to the next section.

These dimensions combine to perform the total epistemic job of the theory. We say that a theory yields knowledge when it depicts symbolically the physical world, when it allows us to solve problems by suitably transforming this symbolic picture, and when it makes sense of what happens. In this framework, the Whewell thesis can be read as the claim that all knowledge content pertains to a few elements that constitute the theoretical core, and the three dimensions are either empty or represented by those elements. In particular, the Whewell thesis amounts to claiming that some fundamental propositions (e.g., the laws of dynamics and of gravitation) exhaust the representational and explanatory dimensions and that their symbolic transformations are epistemically neutral. In the remainder of this chapter I want to argue for the opposite claim, i.e., (1) each dimension is represented by epistemically active elements and (2) elements belonging to one dimension also can contribute epistemically to the other two dimensions. Note that claim (2) can be also formulated by saying that although we can identify theoretical elements that eminently play a representational, transformational, or explanatory role, in effect they also affect the other dimensions. In other words, I argue that the three dimensions *cooperate epistemically* and there are no principled epistemic asymmetries of the kind held by the Whewell thesis.

1.6.2 Transformation, Representation, and Explanation

Let us begin with the long neglected transformational dimension. It is represented by mathematical methods, formal techniques, and other kinds of practices to manipulate symbols. They have traditionally been considered epistemically neutral because, so the argument goes, they just make an implicit knowledge content explicit. My contention is that symbolic practices contribute to a form of representational knowledge and to argue for this claim I elaborate on point (4) above.[13] Symbolic practices are tools, and tools are bridges between us and their object. As such, tools must match both our cognitive abilities and the set of symbolic relations they have to manipulate. In so doing, they might convey a certain amount of knowledge about our epistemic status, as well as the symbolic codifications. More specifically, they provide us with a general knowledge about how the world should be for the symbolic practices to apply successfully. When we study a special problem, we combine a symbolic codification and symbolic practices to obtain prediction. The former informs us about specific physical items, the furniture of the universe as I called it above, whereas the latter gives us knowledge on some general feature of the world. Thus, symbolic practices contain an *epistemic surplus* that partly affects the way in which we think about the problem itself. To put it in a slogan, the fact that I am successfully using a hammer to solve a problem tells me something about the world beyond the problem, namely that it might be seen as made of nails.

Let us pick Whewell's example. For sure, Newton's law of gravitation represents the celestial phenomena and stores a knowledge content to be unfolded. But when we turn this law into equations to calculate planetary motion, we have to make a number of important steps. We have to codify this motion in terms of orbital elements, we need to choose a suitable coordinates system, use symmetry considerations to simplify the expressions, find integrals of motion, and likely elaborate approximate solutions in trigonometric series. The solution of the special problem contains some sort of knowledge that, as Cartwright argues, is not present beforehand in the law of gravitation. Insofar as a theory deals with specific problems, it is constituted with many *centers* in which symbolic codifications and practices are aligned and cooperate epistemically to solve a problem and to constitute a portion of the theory.

Note, however, that, although a symbolic practice says something about the object manipulated, what it actually says may be a matter of contention. Probabilistic methods are a case in point. The successful application of probabilistic techniques can be read as revealing something about the 'randomness' of the world or about our incomplete knowledge of the relevant aspects of the problem. The direction to take depends, ultimately, on the theoretician: she decides the structure of the theory. However, any choice made affects the overall epistemic cooperation of the theoretical dimensions and can generate internal tensions. In Chaps. 3 and 4 we will see how these elements play out in Planck's formulation of the second argument for irreversibility and how the quantum brought to a climax the tensions of this argument.

Before looking at the role of the explanatory dimensions, I have to clarify what I mean by explanation. The philosophical literature on explanation is even more gigantic than on theories, so I will not review it here. For my present aim, I only need a very sketchy proposal that summarizes the indispensable conditions for an explanation of a physical phenomenon. I submit that these conditions are:

EX1 A set of causes or reasons of a certain kind.
EX2 A complete mechanism connecting the causes with the actual phenomenon.

---

[13] I also use some insights of Tom Nickles and of Gerd Gigerenzer and Thomas Sturm in (Gigerenzer and Sturm 2007).

To be sure, these are the conditions for an explanation, namely a justification of why and how the phenomenon came about. But in science one looks for *good* explanations, i.e., explanations that conform to some historical, social, and cultural standards shared by the community. I therefore add a third condition:

EX3 To be acceptable by a given community, the explanation must satisfy the historical, social, and cultural requisites of the community.

A physical explanation is then an *epistemic story* that satisfies conditions (EX1)–(EX3), i.e., it provides the causes of the phenomenon, connects the causes with the phenomenon by means of an articulated mechanism, and is a good story according to some historical, social, and cultural norms. Being a story, an explanation is a story *of a certain kind*, i.e., it is articulated according to an *epistemic trope*. Much like in the case of literary narratives, an epistemic trope establishes the kind of explanation because it defines what kind of *explanantia* and argumentative patterns can be used in the epistemic story. The explanatory dimension of a theory is constituted by epistemic tropes. They can be methodological rules, cultural standards internalized by the theory, or even special interpretations of its constituents. Mechanical explanations admit stories arranging forces, masses, and other mechanical quantities in argumentative patterns that make use of the laws of dynamics and certain mathematical techniques. Analogously, evolutionary stories are made of arguments based on the common ancestors and the selection of the fittest. To take another example I discuss thoroughly later on, in Planck's radiation theory, the strict view of the second law of thermodynamics clearly works as an epistemic trope. It determines the acceptable argumentative patterns and explanantia to be used to explain the irreversibility of the cavity radiation. Note that epistemic tropes characterize a *type* of story that can be realized in different ways. For instance, after Boltzmann's criticism, Planck introduced a new argument and new theoretical ingredients to explain the cavity radiation. Both arguments were realizations of the same epistemic trope.

Epistemic tropes are essentially elements of selection and organization of the epistemic resources. Their contribution to the transformational and representational dimensions consists in aligning the several centers of symbolic codifications and practices in order to produce satisfactory epistemic stories. Symbolic codifications and practices can solve specific problems and obtain astonishing results, but they do not contain their own goals. They need to be organized in preconceived argumentative patterns to produce explanations. Thus, the overall explanation consists of multiple individual steps that require local alignments of the representational and transformational dimensions.

However, these dimensions contribute the ingredient of the explanation. But they do more than just this. As we have seen, we look for a good explanation as given by using the 'right' explanantia and mathematical techniques. Representational and transformational dimensions contribute (positively or negatively) to requisite (EX3) and make an explanation more or less socially, historically, and culturally acceptable. What makes a community accept or reject an explanation is precisely the use of familiar concepts, models, laws, and methods to explain phenomena.

1.6.3 Complex Morphology

In the foregoing analysis I have presented a general framework to conceptualize the production and functioning of theories. I suggest that we have to distinguish between representational, transformational, and explanatory dimensions of a theory. Although these dimensions are represented by different theoretical ingredients, none of them is epistemically neutral and in fact, they all cooperate epistemically to the production of knowledge. My discussion of the epistemic relations between the dimensions has no ambition to be exhaustive. I expect this discussion to be enlarged by further insights from cognitive sciences, cultural studies, sociology, philosophy of mind, philosophy of science, epistemology, and history of science. My goal is to formulate a workspace to make these fields interact to create a fuller picture of scientific theories. My foremost interest in this book is to explore the implications of this workspace for historiography of theories. Can we formulate an approach able to illuminate how physical theories produce knowledge consistent with—and preferably stemming from—their being historically, socially, and culturally situated phenomena?

The first, clear implication is that my workspace entails a much more complex morphology of physical theories. Philosophy of science has traditionally endorsed a morphology based on an epistemically rich theoretical core and a dispensable protective belt. In my picture, one cannot pinpoint a single core from which the epistemic essence of the theory emanates. The several centers of cooperation between symbolic codifications and practices can be arranged differently and play different roles at different times. The epistemic tropes determine the general character of the explanation, but it can be realized by means of different argumentative patterns arranged in different manners. Which elements play representational, transformational, or explanatory roles and how they relate must be established by case-specific historical analysis. A theory appears as a multidimensional object of cooperating codifications and practices whose detailed morphology is complex and cannot be ascertained a priori.

Let us anticipate another instance that I discuss in details below. A key passage in Planck's radiation theory is the introduction of the hypothesis of natural radiation. On the one hand this hypothesis can be considered as a symbolic codification because it yields a picture of how the cavity field behaves. On the other hand, its epistemic content is not only representational. If we open the black box of the theory, we realize that it is strongly motivated by Planck's use of Fourier series to manipulate the field quantities and, of course, by the requirements of his second argument for irreversibility. Although Planck brings about the hypothesis as a choice grounded in simplicity and good sense, a closer analysis of the cooperation between representational, transformational, and explanatory dimensions reveals a more articulate epistemic architecture and a precise strategy.

Thus, my workspace for theories goes hand in hand with a historiographical approach that pays attention to the way in which the three dimensions are represented, how they interact, what kind of epistemic surplus is carried by specific mathematical techniques, what epistemic tropes are used, what sort of argumentative patterns realize them, and so on. I call this methodology Historical-Morphological-Analysis (HMA). It focuses upon the morphology of the theory under study in order to unveil which elements make up the theory and how they came together to create its epistemic story.

There is a consolidated tradition that views scientific theories as trees. After planting a seed that contains the whole theory *in potentia*, one has to wait for it to emerge. The HMA depicts a theory as a turbulent, ever-changing, and articulated entity. A theoretician does not simply cultivate the fruits borne by one genial idea, but continuously balances the epistemic cooperation between the three dimensions of the theory. There might be many ways to carry out this balance and any good theoretician must use all her experience, intuition, good sense, and ability to elaborate a consistent theory. When Lakatos describes the way in which a protective belt is constructed, he argues that the hierarchy of the anomalies is established following an order and "the order is usually decided in the theoretician's cabinet" (Lakatos 1978, p. 49). The very image of a cabinet embodies the positivist ideal of a theory: crystalline, atemporal, well-bounded order. I do not see physical theories as well-defined objects that can be given, delimited, and selected.[14] Theories are embodied in a web of texts and practices, and therefore they have porous boundaries and changing morphologies. I see theories as systems of aligned cognitive resources to produce and manipulate symbols, and I see these cognitive resources as inherently belonging to a historically and culturally situated community. Hence, I see the historical and philosophical analysis of theories as an attempt at entering the theoretician's workshop: a complex and ever-evolving collection of concepts, methods, tricks, assumptions, models, tools of various sorts, experience, tradition, and goals.

1.7 Structure of the Book

The only way to convincingly argue in favor of the relevance of the framework proposed above is to use it in concrete historical research. This is what I try to do in this book. The question to answer now is: why Planck? Well, this is actually the combination of two other questions.

First, why only Planck *and not others*? One might be tempted to claim that a philosophical approach to theories needs to be tested on as many case studies as possible. I agree with this, with a pragmatist caveat: as many case studies as possible, but not all at the same time. I believe that HMA performs best in the careful, detailed analysis of one case at the time, rather than in quickly painting long-term developing lines with a broad brush.

Second, why Planck *instead of others*? A very disappointing answer immediately comes to mind: Planck's radiation theory—and its relations with Boltzmann's statistical mechanics—has been a lifelong interest of mine. The framework discussed

---

[14]On this point see also (Kaiser 2005, pp. 377–387).

above has emerged as a way to make sense of what I have seen in Planck's theory. But there is another reason that goes deeper than that. Planck's radiation theory is one of the few cases in which we can follow very closely the development of an important physical theory. True, we do not possess Planck's research notes and his personal correspondence. But the papers he published between 1897 and 1900 were primarily communications read at the Prussian Academy of Science or the German Physical Society. Planck considered them works-in-progress, and their analysis allows us to map the unfolding of Planck's thoughts, his deadlocks, his strategic withdrawals, and his changes of tack. We have therefore a rich material to disassemble and to dissect in order to understand how the different parts of a theory come together, relate to each other, and evolve in the course of time.

The book is divided into four parts. Chapter 2 discusses the origins of the problem of heat radiation and summarizes the results Planck had at his disposal when he began his program. This chapter tries to spell out the dominant theoretical tradition in order to assess the originality of Planck's program. Being at the boundary between thermodynamics and electromagnetism, the case of heat radiation rapidly became the ideal site to apply pure thermodynamics and Maxwell's electromagnetic theory, as well as kinetic approaches. At the same time, experiments also played a prime role. Its generality notwithstanding, the study of heat radiation benefitted from the migration of conceptual and experimental techniques developed for handling cognate problems such as dispersion and fluorescence. The work of Wilhelm Wien, discussed in Sect. 2.3, was the ripest fruit of this tradition. To arrive at the radiation law that carries his name, Wien combined resources from thermodynamics and electromagnetism plus specific kinetic hypotheses on the microscopic origin of heat radiation. In his program, the question about the form of the radiation law was always coupled with the question of giving a model of interaction between light and matter that might be generalizable to more complex cases.

The second part, which occupies Chap. 3, deals with Planck's radiation theory. In this part, the discussion of Planck's Pentalogy in Sect. 3.2 occupies the most relevant portion in my argument. Planck's initial hope was that the interaction between a field and a Hertzian resonator had the character of absolute irreversibility. The equations and boundary conditions of this electromagnetic process, so Planck thought, are such to rule out any direct reversal (Sects. 3.2.1–3.2.4). This hope was destroyed by Boltzmann, who showed incontrovertibly that Planck had badly overlooked the reversal of the emitted wave. Contrary to Kuhn and other historians, I argue that Boltzmann's criticism did not force Planck to an unconditional acceptance of kinetic theory. Planck recognized that some insights of the kinetic theory were required by his approach, but he tried to adapt very selectively the conceptual resources of kinetic theory to his scheme. I discuss this important process of adaption in detail especially as far as the hypothesis of natural radiation is concerned (Sect. 3.2.6). In a nutshell, my claim is that Planck took his second argument for irreversibility from Boltzmann's H-theorem, but read the mathematical structure differently. For him, that argument revealed something about the relation between macroscopic quantities, not about the microworld. The key for this new reading is the hypothesis of natural radiation. The second argument led Planck to bind the fate of his whole program to a very specific form of the radiation law, that is Wien's law (Sects. 3.2.7–3.2.8). Thus, in hindsight, Boltzmann's criticisms are the condition of possibility for the quantum revolution because initiated that process of 'rigidifying' of Planck's theory that eventually led to the break down.

The third part describes how Planck's theory became increasingly rigid until it could no longer react to the experimental challenges. In the attempt at filling the gaps of the Pentalogy, Planck conceived a new thermodynamic argument that enchained his program to a very specific formula for the second derivative of the entropy (Sect. 4.2). When new experimental data showed that the right distribution law called for a different second derivative, his program was again in trouble. In Sect. 4.5, I discuss the introduction of the quantum hypothesis ensuing this crisis. I carefully describe Boltzmann's combinatorics as the main resource for Planck (Sect. 4.5.1), but at the same time I stress that Planck adapted Boltzmann's procedure to his own problem following different insights (Sect. 4.5.2). There are two important consequences that I stress in Sects. 4.5.2 and 4.6. First, Planck discovered that his use of the statistical procedure was compatible with two combinatorial models. This led him to suspend the judgment on the physical reality of the quantization of energy. Second, in Planck's original scheme, the quantum hypothesis was totally separate not only from the rest of the theory, but also from the physical knowledge of the time. Planck's adaption changed the structure of the statistical procedure, and his opportunistic stance left the role of the new assumption undefined.

Finally, in the fourth part I describe the attempts at understanding the relations between the quantum hypothesis and the available physical knowledge (Sect. 4.7). I focus particularly upon the way in which the physical community tried to spell out the connections between Planck's statistical argument and the tradition of statistical physics.

References

Achinstein P (1963) Theoretical terms and partial interpretation. Br J Philos Sci 14:89–105

Achinstein P (1968) Concepts of science. John Hopkins Press, Baltimore

Arabatzis T, Schickore J (2012) Introduction: ways of integrating history and philosophy of science. Perspect Sci 20:395–408

Bailer-Jones DM (2009) Scientific models in philosophy of science. University of Pittsburgh Press, Pittsburgh

Barnes B (1974) Scientific knowledge and sociological theory. Routledge, Boston

Beth EW (1948) Natuurphilosophie. Noorduyn, Gorinchem

Beth EW (1960) Semantics of physical theories. Synthese 12:172–175

Burian RM (1977) More than a marriage of convenience: on the inextricability of history and philosophy of science. Philos Sci 44:1–42

Carnap R (1936–1937) Testability and meaning. Philos Sci 3, 4:420–468; 1–40

Carnap R (1956) The methodological character of theoretical concepts. In: Feigl H, Scriven M (eds) The foundations of science and the concepts of psychology and psychoanalysis, minnesota studies in the philosophy of science, vol 1. University of Minnesota Press, Minneapolis, pp 38–76

Cartwright N (1983) How the laws of physics lie. Oxford University Press, Oxford

Craig W (1953) On axiomatizability with a system. J Symbolic Logic 18:30–32

Da Costa NCA, French S (1998) Models, theories, and structures: thirty years on. Philos Sci 67: 116–127

Galison P, Warwick A (1998) Introduction: cultures of theory. Stud Hist Philos Mod Phys 29(3): 287–294

Giere RN (1973) History and philosophy of science: intimate relation or marriage of convenience? Bri J Philos Sci 24:282–297

Giere RN (1988) Explaining Science: a cognitive approach. University of Chicago Press, Chicago

Giere RN (2001) Theories. In: Newton-Smith WH (ed) Blackwell companion to the philosophy of science. Blackwell, Malden, MA, pp 515–524

Giere RN, Richardson A (eds) (1996) Origins of logical empiricism, minnesota studies in the philosophy of science, vol XVI. University of Minnesota Press, Minneapolis

Gigerenzer G, Sturm T (2007) Tools = Theories = Data? On some circular dynamics in cognitive science. In: Ash M, Sturm T (eds) Psychology's territory. Historical and contemporary perspectives from different disciplines, Lawrence Erlbaum Associates, Mahwah, NJ, pp 305–342

Hempel CG (1958) Theoretician's Dilemma. A Study in the logic of theory construction. In: Feigl H, Scriven M, Maxwell G (eds) Concepts, theories, and the mind-body problem, minnesota studies in the philosophy of science, vol 2. University of Minnesota Press, Minneapolis, pp 37–98

Kaiser D (2005) Drawing theories apart. University of Chicago Press, Chicago, The Dispersion of Feynman Diagrams in Postwar Physics

Krüger L (1982) History and philosophy of science—a marriage for the sake of reason. In: Cohen LJ, Los J, Pfeiffer H, Podewski KP (eds) Proceedings of the VI international congress for logic, methodology and philosophy of science, Hannover 1979. North Holland, Amsterdam, pp 108–112

Lakatos I (1978) The methodology of scientific research programmes, philosophical papers, vol I. Cambridge University Press, Cambridge

Laudan L (1977) Progress and its problems. University of California Press, Berkeley

Meheus J, Nickles T (eds) (2009) Models of discovery and creativity. Springer, New York

Mormann T (2007) The structure of scientific theories in logical empiricism. In: Richardson A, Uebel T (eds) The cambridge companion to logical empiricism. Cambridge University Press, Cambridge, pp 136–162

Morrison M (2000) Unifying scientific theories: physical concepts and mathematical structures. Cambridge University Press, Cambridge

Morrison M, Morgan MS (1999) Models as mediating instruments. In: Morrison M, Morgan MS (eds) Models as mediators. Cambridge University Press, Cambridge, pp 10–37

Nickles T (1978) Scientific discovery and the future of philosophy of science. In: Nickles T (ed) Scientific discovery, logic, and rationality. Kluwer Academic, Dordrecht, pp 1–59

Nickles T (2001) Discovery. In: Newton-Smith WH (ed) Blackwell companion to the philosophy of science, Blackwell, pp 85–96

Putnam H (1962) What theories are not. In: Nagel E, Suppes P, Tarski A (eds) Logic, methodology, and philosophy of science: Proceedings of the 1960 international congress. Stanford University Press, Stanford, pp 240–251

Reichenbach H (1938) Experience and prediction. University of Chicago Press, Chicago

Shapere D (1984) Reason and the search for knowledge. Reidel, Dordrecht

Shapere D (1987) Method in the philosophy of science and epistemology: how to inquire about inquiry and knowledge. In: Nersessian NJ (ed) The process of science. Contemporary philosophical Approaches to Understanding Scientific Practice, Martinus Nijhoff Publishers, Dordrecht, pp 1–39

Stadler F (2007) The Vienna circle. Context, profile, and development. In: Richardson A, Uebel T (eds) The Cambridge companion to logical empiricism. Cambridge University Press, Cambridges, pp 13–40

Suppe F (1977) The search for philosophical understanding of scientific theories. In: Suppe F (ed) The structure of scientific theories. University of Illinois Press, Chicago, pp 3–232

Suppe F (1989) The semantic conception of theories and scientific realism. University of Illinois Press, Chicago

Suppes P (1967) What is a scientific theory? In: Morgenbesser S (ed) Philosophy of science today. Basic Books, New York, pp 55–67

Suppes P (1969a) A comparison of the meaning and use of models in the mathematical and empirical sciences. In: Studies in the methodology and foundations of science. Selected Papers from 1951 to 1969. Springer, New York, pp 10–23

Suppes P (1969b) Models of data. In: Studies in the methodology and foundations of science. Selected Papers from (1951) to 1969. Springer, New York, pp 24–35

Van Fraassen B (1970) On the extension of Beth's semantics of physical theories. Philos Sci 37: 325–339

Van Fraassen B (1980) The Scientific image. Oxford University Press, Oxford

Van Fraassen B (1987) The Semantic approach to scientific theories. In: Nersessian NJ (ed) The process of science. Martinus Nijhoff Publishers, Dordrecht, Contemporary Philosophical Approaches to Understanding Scientific Practice, pp 105–124

Whewell W (1837) Astronomy and general physics considered with reference to natural theology. William Pickering, London

Chapter 2
The Problem of Heat Radiation

Abstract This chapter summarizes the problem of heat radiation in the second half of the nineteenth century. For most physicists, this problem amounted to finding the explicit form of the radiation law. In the first phase, experimental research and general thermodynamical arguments imposed some constraints on the form of this law. One of the great conundrums of the final decades of the nineteenth century was to discover a plausible derivation of the exponential term revealed by the experiments. Here, I pay special attention to Wien's research program. Wien combined electromagnetic theory, kinetic theory, and thermodynamics in a very creative—and sometimes opportunist—manner. More importantly, for Wien the black-body problem was a window on the study of more intricate forms of interaction between radiation and matter. Planck's program, as we will see in the next chapters, had a totally different agenda.

Keywords Heat radiation · Black-body · Thermodynamics · Electromagnetism · Wien

2.1 The Origins

In the second half of the nineteenth century, physicists began to be increasingly concerned with the radiations emitted by bodies. In the vast phenomenology of the induced emissions, the so-called heat radiation occupied a sort of privileged position. The reason was that such radiation was considered to be the basis for more complex processes such as fluorescence and phosphorescence. Although heat radiation depends on the temperature of the body and, in general, on its chemical and physical nature, it turned out that its essential features could be represented in a very simple way.

Gustav Robert Kirchhoff (1824–1887) was the first to give a full-fledged theoretical treatment of heat radiation in 1860. By combining thermodynamic arguments and phenomenological considerations, he clarified the general properties of thermal emission and absorption. The most original conceptual tool introduced by Kirchhoff in his research was the notion of an ideal body, "which absorb[s] all the rays that fall upon [it] at infinitesimal density" (Kirchhoff 1860, p. 277). Because no radiation

© The Author(s) 2015

M. Badino, *The Bumpy Road*, SpringerBriefs in History
of Science and Technology, DOI 10.1007/978-3-319-20031-6_2

is reflected, this object was called a black-body. Later in the same paper, Kirchhoff explained how to experimentally construct a black-body:[1]

> If a cavity is surrounded by bodies at the same temperature and no ray can penetrate these bodies, then each bundle of rays in this cavity possesses the same quality and intensity as though it came from a perfectly black-body at the same temperature, i. e. independent of the particularities and form of the body and constrained by the temperature only. (Kirchhoff 1860, p. 300)

More importantly, Kirchhoff also showed that the ratio between emissive power $e(\lambda, T)$ and absorption power of a body $a(\lambda, T)$ is a universal function of the wavelength and the temperature:[2]

$$\frac{e(\lambda, T)}{a(\lambda, T)} = F(\lambda, T), \tag{2.1}$$

independently of the form of the bodies or the substance they are made of. Kirchhoff's analysis had two important consequences. Firstly, it entailed that each wavelength can be treated individually as far as its thermodynamical properties are concerned: the radiation is an ensemble of independent components of different wavelengths. Secondly, because for a black-body $a(\lambda, T) = 1$, the emissive power of such a body yielded the sought-for universal function. Kirchhoff concluded that "to find this function is a task of the utmost importance" (Kirchhoff 1860, p. 292).[3]

These remarkable accomplishments encouraged further experimental and theoretical research on the properties of the black-body. The experimental investigation benefitted very much from the introduction of the interferometer and the bolometer in the 1880s.[4] A massive collection of precise measurements ranging from the visible to the infrared region and carried out by means of techniques used in the study of optical dispersion and fluorescence allowed for the determination of some general characteristics of the distribution function $F(\lambda, T)$. One of the first properties to be ascertained was the 'displacement' of the maximum of energy. If one plots the wavelength against the energy at a certain temperature, one discovers that the wavelength $\lambda_{\max}$ corresponding to the maximum energy changes with the temperature because the product $\lambda_{\max} T$ is a constant (Crova 1880; Langley 1886a, b).

A second important property, also firmly established by bolometric experiments, was that the irradiated energy decreases toward both extremes of the spectrum, and the decreasing becomes faster toward the violet limit. These two pieces of information suggested that the correct distribution law had a bell-like form typical of the Gaussian or Maxwellian curve. From the analytical standpoint, this entailed that the universal function $F(\lambda, T)$ contained a suitably weighted exponential term. Further information on the distribution function came from the analysis of the total energy irradiated

---

[1] On the history of the black-body see also (Darrigol 1992; Kangro 1970; Kuhn 1978).

[2] The emissive power is the quantity of energy emitted by the body in radiation form in time unit at a given wavelength and temperature; the absorption power for the same wavelength and temperature is the fraction of impinging energy absorbed by the body.

[3] On the complex story of Kirchhoff's law and its disputed proof see (Schirrmacher 2003).

[4] On bolometric experiments see (Loettgers 2003; Lombardi 2003).

by a black-body. In 1879, Josef Stefan guessed from experiments that the energy emitted by bodies was proportional to the fourth power of the temperature. In 1884, Ludwig Boltzmann (1844–1906) gave a rigorous derivation of this result (Boltzmann 1884a, b; Stefan 1879).[5] He drew upon the work of Adolfo Bartoli, who had discovered an interesting analogy between a cavity full of radiation and a thermodynamic system.[6] By arranging black and reflecting cavities it was possible to construct a cyclic process in which heat was transmitted from one body to another. The cavity radiation could be handled by the usual methods of thermodynamics. To remain consistent with the second law, one had to assume that radiation exerted a pressure so that a performance of work would compensate for the transfer of heat. Boltzmann realized that by combining this process with the concept of radiation pressure proposed by James Clerk Maxwell (1831–1879) in his electromagnetic theory, he could derive a relation between energy density and temperature.

According to Maxwell, the pressure exerted by an electromagnetic radiation on a surface depends on the density of radiation impinging perpendicularly on the surface. In a cavity, heat rays blaze along any direction, but if we assume the cavity to be a parallelepiped, on average the quantity of energy hitting a selected surface of the cavity perpendicularly is one third of the density. This was nothing but the application of an argument already common in kinetic theory. If the heat radiation produces pressure, it is easy to get a relation, which has the form of an exact differential, between pressure, temperature, and volume variation. By replacing the pressure with the energy density u_λ, Boltzmann found immediately:

$$\int_0^\infty u_\lambda d\lambda = \int_0^\infty F(\lambda, T)d\lambda \propto T^4. \tag{2.2}$$

Together with the earlier experimental insights, the Stefan-Boltzmann law stimulated bolder guesses on the form of the distribution function. For instance, Heinrich Weber (1843-1912) extrapolated from his and others' experimental results an exponential expression for the energy density that satisfied the essential condition of displacement (Weber 1888). More interesting, though, was the theoretical derivation of the Russian physicist Vladimir Michelson. The key idea was to connect the exponential term to some sort of Maxwell's distribution in the radiation field by assuming that heat radiation was produced "by the complete irregularity of the vibrations of [the] atoms." (Michelson 1887, p. 426). Michelson supposed that atoms in solids can vibrate around equilibrium positions. Contrary to the usual theory of solids, though, he did not assume a restoring force: atoms are supposed to move freely within a sphere centered in the equilibrium position, with their trajectory being the composition of rebounds at the boundary of the sphere and erratic influence of the neighbor

---

[5]The Stefan-Boltzmann law enjoyed almost immediate experimental confirmation; although Heinrich Weber pointed out deviations at higher temperature (Weber 1888), it was found perfectly correct up to 1535 K by the end of the century (Lummer and Pringsheim 1897).

[6]On Bartoli's work on radiant heat see (Carazza and Kragh 1989).

atoms. This peculiar assumption allowed Michelson to introduce equiprobability (of the positions in the sphere and of the directions of motion) and to show that the most common trajectory passes through the center as a diameter.

To this already disputable derivation, Michelson added two further assumptions. Firstly, he assumed that an atom vibrating along a diametrical trajectory gives off radiation whose period depends on the velocity v of the atom and on the radius ρ of the sphere according to the relation $\tau = 4\rho/v$. If the velocities are distributed according to Maxwell's law, one can find the number of atoms giving off radiation of a certain period simply by replacing v with τ in the usual Maxwell distribution.

Secondly, the intensity of the emitted radiation must depend on the number of atoms vibrating with the corresponding period, on a function of the period itself, and on a function of the absolute temperature. Michelson offered no justification for these suppositions. Imposing the validity of the Stefan-Boltzmann law, the energy intensity as a function of the wavelength becomes:

$$I_\lambda d\lambda = BT^{3/2}e^{-\frac{c}{\lambda^2 T}}\lambda^{-6}d\lambda, \qquad (2.3)$$

where B, c are constants. Michelson's law fitted the bolometric data reasonably well, although it had an incorrect argument in the exponential term and an incorrect dependence on the wavelength. The important point, however, was the connection between the exponential term and the kinetic theory of gas: this connection appeared to be the most promising way to derive the distribution law.

2.2 Entering Electromagnetism

Heinrich Hertz's (1857–1894) pioneering experiments on the properties of electromagnetic waves boosted further research on heat radiation. Treating thermal radiation as electromagnetic waves enabled physicists to use Maxwell's powerful formalism to understand the nature of heat radiation and to clarify the difficult interaction between light and matter. I discuss extensively the second point in the next sections. Here, I focus upon the relation between the so-called Hertzian waves (in the region of the radio waves) and the heat waves (located in the infrared region). Experimenters used basically the same model deployed to explain optical dispersion and fluorescence: light interacts with matter (atoms or molecules) by co-vibrating with it. In the case of radio wave, it was supposed that rotation was also somehow involved. This model left open the problem of what kind of matter could interact with what kind of light: were Hertzian waves produced by atoms, molecules, or even heavy molecules? Perhaps more importantly, the question of what kind of light produced thermal effects was also on the table. It is important to realize the connection between these experiments and the issue of heat radiation. For most physicists, the problem of deriving the black-body radiation law was tightly related to the problem of understanding the interaction between light and matter down to a very specific experimental description. The very abstract representation provided by the black-body model had to be coupled with a detailed analysis of the constituents of matter and light. Wilhelm Wien (1864–1928), whose research program I discuss in the following section, drew heavily on this tradition, while Planck, meaningfully, ignored it entirely.

One notable example of these experimental studies at the beginning of the 1890s is Viktor Bjerknes's work on the propagation of Hertzian waves in grids and metal lattices. From his research, Bjerkens concluded that while optical phenomena involved the interaction between light and individual molecules, as soon as one moves to the regime of Hertzian waves, the interaction concerns whole groups of molecules: "between the Hertzian and the optical oscillation there must be a transition in which the molecules stop acting together and start doing it individually" (Bjerknes 1893, p. 604). This result cautioned against extending the usual optical formalism to the problem of the interaction of matter and radiation beyond the visible region. Wien elaborated Bjerknes's results on the behavior of light through grids and concluded, on the ground of thermodynamic considerations, that there is an upper limit for the light that can make up heat radiation: "we have to accept that the upper limit of the wavelengths that can be produced by heat lies between the Hertzian oscillations and the infrared rays so far observed" (Wien 1893a, p. 636). For long wavelengths it was difficult to assume resonance phenomena between light and matter (Garbasso and Aschkinass 1894), but in the regime described by Wien, the usual formal machinery of dispersion theory can be used, as was confirmed by Heinrich Rubens (1865–1922) slightly afterwards (Rubens 1894).

These investigations fueled the study on the heat radiation in that they provided a map of the relations between thermal phenomena, optical phenomena, and the new Hertzian waves. The general character of the heat radiation notwithstanding, the physicists coupled the problem of the radiation law together with questions about the theoretical and the experimental details of the interaction between light and matter.

2.3 Wien's Research Program

Planck's program was by no means the sole effort to tackle the problem of heat radiation. It is a historically relevant question to ask preliminarily what characterizes Planck's approach in comparison to similar attempts. An answer to this question will help our investigations in two ways. Firstly, it will show that Planck found many of the concepts, techniques, and formal tools disseminated in the works of the physicists engaged in similar questions. Planck's program was deeply rooted in classical physics. At the same time, and this is the second point, classical physics was made up of many different theoretical traditions: a common disciplinary backdrop such as radiation theory might manifest alternative—and sometimes competing—sets of concepts, priorities, argumentative patterns, and techniques along with different sets of questions. Briefly said, representational, transformational, and explanatory dimensions varied remarkably in the different approaches to the black-body problem. We do not have to take for granted that Planck's final goal was the same as that of his

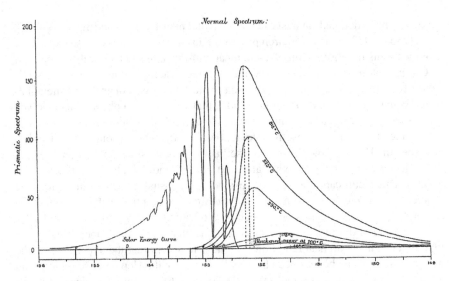

Fig. 2.1 The displacement of the maximum of energy (from Langley's paper)

colleagues, who were working on the same topic. To analyze Planck's competitors means to place Planck's program in the right theoretical tradition.

The best term of comparison for this task is the research program on heat radiation that Wilhelm Wien pursued during the first half of the 1890s. Drawing on the battery of experimental results on the properties of the black-body curve and on the nature of the interaction between light and matter, Wien set for himself a clear goal: to determine the explicit form of the radiation law. This mission had to be accomplished in two steps. The first step consisted of giving a thermodynamic foundation to the still shaky knowledge about the general characteristics of the radiation law. This foundation would also yield the outlines of a thermodynamics of heat radiation. In the second step, one had to supplement the previous considerations with a model of interaction between light and matter that was (1) realistic as far as the experimental knowledge was concerned and (2) able to provide a specific form of the radiation law, in particular the exponential term (Fig. 2.1).

Wien started by giving a thermodynamic argument for the already observed displacement of the maximum of the energy intensity. The basic idea of his approach is that the heat radiation can be treated as a thermodynamic system whose characteristic variables are wavelength, temperature, pressure, volume, and energy. This entails that thermodynamic transformations (temperature variations) can be combined with mechanical one (volume variations) and, more importantly, electrodynamic ones (wavelength variations using the Doppler effect) in order to obtain information on the way in which the radiation changes along with the characteristic variables. Wien assumed three hollow cavities full of radiation like in Fig. 2.2 (Wien 1893b); the picture is from (Wien 1894, p. 134).

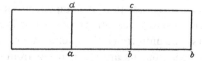

Fig. 2.2 An arrangement of three hollow cavities full of radiation; ad and bc are movable pistons whose surface diffuses radiation in the cavity; the two outward walls are black-bodies

The partitions ad and bc between the cavities are movable, and they can also be taken out to let the radiation pass from a cavity to another. By means of this arrangement it becomes possible to study the variation of the spectral energy density. An adiabatic movement of the partitions produces a change in the energy density (volume variation), as well as a wavelength variation due to the Doppler effect. If these two concomitant changes are calculated independently, their ratio turns out to be $u/u_0 = \lambda_0^4/\lambda^4$, where u_0, λ_0 are the value of the energy density and the wavelength, respectively, before the adiabatic process. But the Stefan- Boltzmann law states that $u/u_0 = T_0^4/T^4$; therefore Wien can immediately derive the law:

$$\lambda T = \lambda_0 T_0. \tag{2.4}$$

This general relation holds true for all wavelengths and temperatures: applied to the wavelength with the maximum energy it expresses the 'displacement' effect noticed by the experimenters. The combination of the Stefan-Boltzmann law and the displacement law constrains remarkably the form of the radiation law. For if we know the emissive power $e(\lambda_0)$ as a function of the wavelength, then we can calculate it after the transformations mentioned above by simply replacing the initial wavelength in the argument, that is $e\left(\frac{\lambda T}{T_0}\right)$. This entails that the emissive power (or the energy density or energy intensity) is a function of the product λT. If we now impose the condition that the integral of the emissive power over all wavelengths is proportional to T^4, it turns out that:[7]

$$e(\lambda, T) = \text{const} \cdot T^5 F(\lambda T) = \frac{\text{const}}{\lambda^5} F(\lambda T). \tag{2.5}$$

The following year, Wien furthered his analysis of the heat radiation by establishing an analogy with kinetic theory. In the experimental investigations, the temperature of the radiation was customarily reduced to the temperature of the source, usually a solid body. However, Wien noticed that if a cavity full of radiation is a genuine thermal system, it must be possible to assign a temperature, and even an entropy, to it. To do this, the black-body comes in handy as an ideal thermometer. Here, for the first time, Kirchhoff's ideal black-body is defined in terms of the state of thermal equilibrium: "the radiation of a black-body is henceforth the state of stable equilibrium into which any radiation of different property will transform *autonomously*" (Wien

---

[7]See (Wien 1894, pp. 158–159); see also (Wien 1909, p. 298).

1894, p. 133). For the black-body radiation, all components (at different wavelength, polarization state, and direction of propagation) have the same temperature, although the energy density differs. The uniformity of the temperature, Wien noticed, derives from the fact that "these components are independent from one another, because we can produce a radiation that contains only one component." This suggests that a cavity radiation is analogous to a mixture of gases of dissimilar chemical species. In an arbitrary state, these components have different temperatures and different energies: the black-body state is reached when the temperature becomes the same for all components and the energy density assumes the simple form required by Kirchhoff's law. To describe how energy can possibly pass from one wavelength to another, the methods of thermodynamics are not enough, and one must introduce an assumption on the microscopic mechanism of interaction.

At this stage, however, Wien is interested in the concept of temperature. Using the black-body radiation as a thermometer, he defines the temperature of a component in an arbitrary radiation state as the temperature that the same component of a black-body radiation would have if it had the same energy density. This definition of temperature is perfectly unequivocal because the black-body radiation curves do not intersect each other, and therefore two different values of the energy density must necessarily correspond to different temperatures.

In the second part of the paper, Wien comes back to electromagnetism: using Maxwell's equations, it is possible to better understand the essence of the cavity radiation. Interestingly, Wien deploys—and it is the first time ever— Hertz's spherical solutions of Maxwell's equations, which were to become Planck's starting point. By means of Hertz's solutions, Wien calculates the Poynting vector and then makes a more accurate evaluation of the work performed by volume increase. But it is the microscopic consequences that interest Wien most. The individual oscillations are perfectly compatible with the laws of mechanics; therefore, if we were able to act on them singularly with a suitable mirror, we could produce violations of the second law of thermodynamics: one can construct an electromagnetic version of Maxwell's Demon.[8] Wien is not afraid of accepting the analogy with kinetic theory until the last consequence: "just as the second principle holds only so long as one can consider the motion of the molecules merely as a whole, thus it ceases to be valid as soon as one can act upon the individual oscillations generated by the heat" (Wien 1894, p. 151).

Finally, Wien introduces the concept of entropy. In thermodynamics, temperature is defined as the integrating denominator of the quantity of heat. Drawing on his previous work, Wien evaluates the variation of the heat content in case of an expansion of the cavity volume. According to the first principle of thermodynamics, such variation is the sum of the change of the internal energy and the external work. Both quantities can be expressed in terms of the energy density. Applying the formal condition for an integrating denominator, Wien finds a relation between temperature and energy density that, unsurprisingly, coincides with the Stefan-Boltzmann law. Dividing the heat content by the integrating denominator, Wien determines the entropy of the

---

[8]On Maxwell's Demon see Sect. 3.2.2.

radiation $S = S_0 + \frac{4dl}{3\sigma} u^{3/4}$, where dl is the displacement of the partition and σ the constant of the Stefan-Boltzmann law.

At this point, it might be useful to recall Wien's conception of entropy. At the beginning of the 1890s, Wien was particularly impressed by the implications of the work of J. H. Poynting, according to whom one can ascribe to energy quantities that are commonly associated with matter such as localization and motion. We can talk about a flux of energy in the same sense as we talk about a flux of matter. To Wien, this view suggested a possible reconciliation of the two laws of thermodynamics. Indeed, contrary to what happens in the dynamics of fluids, energy has a privileged 'direction'. The entropy principle, in other words, establishes an intrinsic law determining how energy moves in space: "one can express this principle by saying that, given an uneven distribution of its density, the energy moves always so to eliminate this difference" (Wien 1892, p. 724). This interpretation of the concept of entropy strengthens the relation with the principle of least action that Boltzmann and Rudolf Clausius had already stressed in the 1870s (Boltzmann 1866; Clausius 1871). If we confine ourselves to periodic motion, it is possible to show that the requirement that temperature be an integrating denominator for the heat content is equivalent to the Hamilton principle. Wien's conclusion was that:

> [O]ne can apply the entropy principle with certainty only to periodic motions or to motions such that the actual trajectories can be represented by mean values referring to closed trajectories. It is still presupposed that the motion of energy occurs in order to balance out the integrating denominator T. (Wien 1892, p. 726)

The entropy of the radiation is the energy guiding function and depends on macroscopic quantities only. Wien's concepts of entropy and temperature were the best thermodynamics could get without microscopic assumptions. But it was still insufficient to obtain a complete form of the radiation law. In the final pages of the paper, he came back to the analogy with the gas, searching for clues to the solution of the problem:

> When a radiation of a simple color transforms itself into a radiation of mixed colors, the entropy increases and this increase represents a compensation for the possible work obtainable. There is a complete analogy with the entropy increase of a mixture of separate gases. Also in gases the entropy of the mixture is equal to the sum of the entropies of the individual gases as though they were alone in the container. But an irreconcilable difference lies in the fact that the diversities of the gases allow different values of entropy even at the same temperature, while here we have to do with energy that changes the entropy value only through differences in temperature. (Wien 1894, p. 160)

Defining the black-body radiation as the equilibrium state suggested a fairly natural continuation of the argument: supplementing thermodynamic considerations with kinetic-like assumptions to arrive at the radiation law. This was the most obvious way to get the exponential term as Michelson had demonstrated. In 1896, Wien followed precisely this route to complete his program. Here, he puts forward a more general reason why additional microscopic hypotheses of kinetic type were necessary. A purely electromagnetic derivation of the radiation law would call for a mechanism of redistribution of the energies over different wavelengths without performing work or absorbing energy. But, Wien states, "there is at the moment no physical process through which the above mentioned transformation of the colors takes place in a natural way" (Wien 1896, p. 662). For this reason, the radiation law had to be derived by patching experimental information and kinetic hypotheses.

Admittedly, Wien's argument is not much better than Michelson's. He is only more careful about the hypotheses and the general thermodynamic constraints, but the fundamental idea remains the same. He assumes that the wavelength of the radiation emitted by a molecule is a function of its velocity and that the intensity of the radiation at a certain wavelength depends on the number of molecules with the corresponding velocity.[9] In this way one can transform Maxwell's distribution into a distribution over the wavelengths: $u_\lambda d\lambda = F(\lambda)e^{-\frac{f(\lambda)}{T}}d\lambda$. The displacement law and the Stefan-Boltzmann law help determine further the unknown functions. From the former it follows that the energy density must depend on the product λT, and from the latter that it must contain the term λ^{-5}. Putting together these two pieces of information, Wien obtains:

$$u_\lambda d\lambda = \frac{c_1}{\lambda^5}e^{-\frac{c_2}{\lambda T}}d\lambda, \tag{2.6}$$

where c_1, c_2 are two universal constants. This is the longed-for radiation law. Around the same time, Friedrich Paschen arrived at a similar formula by working on the experimental data only (Paschen 1896; paschen 1897). Wien was in contact with Paschen and referred to his law in the paper. More importantly, he was thereby informed that this law fitted very well the experimental results, at least within the range of temperature and wavelength covered by the experimenters. With the derivation of a satisfactory radiation law, Wien's program was finally complete. The apparently shaky derivation was not a big problem: strange as the assumptions could seem, they were rooted in the physical knowledge on the nature of the phenomenon. The study of optical dispersion and fluorescence supported the idea that the radiation emitted by the molecules depended on their velocity, while it was reasonable to think that the intensity depended on their number. For a gas and radiation at relatively short wavelength, the application of Maxwell's distribution was again a natural choice. It was therefore in the microscopic details of the light-matter interaction that Wien looked for the additional hypotheses to supplement the thermodynamic framework.

2.4 Concluding Remarks on the Problem of Heat Radiation

To sum up, when Planck started his own research program, the problem of heat radiation was defined within a framework that comprised the following elements. First, there were some general thermodynamic laws that set many of the essential formal properties of the radiation law. Second, other properties, for instance the presence of an exponential term, followed from the analysis of a huge collection

---

[9]Contrary to Michelson, who had applied his argument to solids, Wien considers only gases because for gases the assumptions were more justifiable.

of experimental data; in addition, these experiments also gave hints on the nature of heat radiation and on the mechanism of interaction between light and matter. The problem of heat radiation was seen as the paradigmatic example of the general question of light-matter interaction: the extreme simplicity of the phenomena made this case the ideal benchmark for an analysis generalizable to more complex cases, such as optical dispersion and fluorescence, that depended instead on the physical features of the bodies (Rubens 1900). The crucial point here is that the simplicity of heat radiation was not a reason to abstain from conjectures on the microscopic nature of the light-matter interaction, as Wien's program demonstrates. Quite the contrary, physicists were trying to make heat radiation the key to treat all instances of light-matter interaction, and to do that, they had to conceive a theory as realistically as possible.

Wien systematically developed this program to its ultimate conclusion. His main aim was to obtain a radiation function, and to achieve that he combined several resources. He used techniques and argumentative patterns both from thermodynamics and from kinetic theory and electromagnetism. The exponential factor suggested a relation with Maxwell's kinetic methods, which in turn led to a certain representation of the microscopic processes of radiation. For Wien, all these ingredients were the representational and the transformational dimensions to be combined into the explanatory dimension of his program. While to Planck, as we will see later on, the generality of heat radiation justified a non-committal attitude toward the microscopic mechanism of production and transformation of the radiation, for Wien the problem was inextricably intertwined with the elaboration of a realistic model. Some years later, he reiterated this point by criticizing Planck's resonators because they "do not change the spectral composition of the radiation and therefore they cannot represent *a model of a radiating body*" (Wien 1909, p. 302, italics added).

The historical question that we must bear in mind while discussing Planck's theory is now to what extent, if any, this criticism actually reveals a failure in Planck's original program. It is doubtlessly true that Planck's arsenal of concepts and techniques partially overlaps Wien's, such that the temptation is strong to ascribe to both identical goals. Planck, however, inscribed his work within a completely different epistemic story, which led him to a very different approach.

References

Bjerknes V (1893) Das Eindringen electrischer Wellen in die Metalle und die electromagnetische Lichttheorie. Annalen der Physik 48:592–605

Boltzmann L (1866) Über die mechanische Bedeutung des zweiten Hauptsatzes der Wärmetheorie. Sitzungsberichte der Akademie der Wissenschaften zu Wien 53:195–220

Boltzmann L (1884a) Ableitung des Stefan'schen Gesetzes betreffend die Abhängigkeit der Wärmestrahlung von der Temperatur aus der elektromagnetiscen Lichttheorie. Annalen der Physik 22:291–294

Boltzmann L (1884b) Über eine von Hrn. Bartoli entdeckte Beziehung der Wärmestrahlung zum zweiten Hauptsatze. Annalen der Physik 22:31–39

Carazza B, Kragh H (1989) Adolfo bartoli and the problem of radiant heat. Ann Sci 46:183–194

Clausius R (1871) On the reduction of the second axiom of the mechanical theory of heat to general mechanical principles. Philos Mag 42:162–181

Crova A (1880) Etude des radiations emises par les corps incandescents. Mesure optique des haute temperatures. Annales de Chimie et de Physique 19:472–550

Darrigol O (1992) From c-numbers to q-numbers. The classical analogy in the history of quantum theory. University of California Press, Berkeley

Garbasso A, Aschkinass E (1894) Über Brechung und Dispersion der Strahlen electrischer Kraft. Annalen der Physik 53:534–541

Kangro H (1970) Vorgeschichte des Planckschen Strahlungsgesetzes. Steiner, Wiesbaden

Kirchhoff GR (1860) Über die Verhältnis zwischen dem Emissionsvermögen und dem Absorptionsvermögen der Körper für Wärme und Licht. Annalen der Physik 109:275–301

Kuhn T (1978) Black-body theory and the quantum discontinuity, 1894–1912. Oxford University Press, Oxford

Langley SP (1886a) Observations on invisible heat-spectra and the recognition of hitherto unmeasured wavelength, made at the allegheny observatory. Philos Mag 21:394–409

Langley SP (1886b) Sur les spectres invisibles. Annales de Chimie et de Physique 9:433–506

Loettgers A (2003) Samuel pierpont langley and his contributions to the empirical basis of black-body radiation. Physics in Perspective 5:262–280

Lombardi AM (2003) The bolometer and the spectro-bolometer as steps towards the black-body spectrum. Nuncius 18:825–840

Lummer O, Pringsheim E (1897) Die Strahlung eines schwarzen Körpers zwischen 100 und 1300° C. Annalen der Physik 63:395–410

Michelson WA (1887) Theoretical essay on the distribution of energy in the spectra of solids. Philos Mag 25:425–435

Paschen F (1896) Über Gesetzmässigkeiten in den Spectren fester Körper. I. Annalen der Physik 58:455–492

Paschen F (1897) Über Gesetzmässigkeiten in den Spectren fester Körper. II. Annalen der Physik 60:662–723

Rubens H (1894) Prüfung der Ketteler-Helmholtz'schen Dispersionsformel. Annalen der Physik 51:267–286

Rubens H (1900) Le spectre infra-rouge. In: Guillaume CE, Poincaré H (eds) Rapports presentes au Congres International de Physique, vol 2. Gauthier-Villars, Paris, pp 141–174

Schirrmacher A (2003) Experimenting theory: the proofs of kirchhoff's radiation law before and after planck. Hist Stud Phys Biol Sci 33:299–335

Stefan J (1879) Über die Beziehung zwischen der Wärmestrahlung und der Temperatur. Sitzungsberichte der Akademie der Wissenschaften zu Wien 79:391–428

Weber HF (1888) Untersuchungen über die Strahlung fester Körper. Sitzungsberichte der Preussischen Akademie der Wissenschaften 2:933–957

Wien W (1892) Über den Begriff der Localisirung der Energie. Annalen der Physik 45:685–728

Wien W (1893a) Die obere Grenze der Wellenlängen, welche in der Wärmestrahlung fester Körper vorkommen können; Folgerungen aus dem zweitem Hauptsatz der Wärmetheorie. Annalen der Physik 49:633–641

Wien W (1893b) Eine neue Beziehung der Strahlung schwarzer Körper zum zweiten Hauptsatz der Wärmetheorie. Sitzungsberichte der Preussischen Akademie der Wissenschaften 1:55–62

Wien W (1894) Temperatur und Entropie der Strahlung. Annalen der Physik 52:132–165

Wien W (1896) Über die Energievertheilung in Emissionsspectrum eines schwarzen Körpers. Annalen der Physik 58:662–669

Wien W (1909) Theorie der Strahlung. In: Sommerfeld A (ed) Encyklopädie der mathematischen Wissenschaften, vol 3, Teubner, Leipzig, pp 282–357

Chapter 3
Planck's Theory of Radiation

Abstract In this chapter I explore Planck's radiation theory from his preliminary studies (1896) through his more mature Pentalogy (1897–1899). Planck viewed the problem of the black-body radiation very differently from Wien and the majority of his contemporaries. In particular, Planck was not primarily interested in deriving a radiation law. Instead, he considered heat radiation as an ideal case to support his strict view of thermal irreversibility. He wanted to prove that electromagnetic radiation in a cavity, when suitably stimulated, reaches irreversibly a form of stable thermal equilibrium. Initially, Planck thought that this statement could be demonstrated as a consequence of the electromagnetic features of the problem. Boltzmann jumped in and showed that this could not possibly be the case. In the second part of the Pentalogy, Planck changed strategy. He modified the morphology of his theory to accommodate new resources and gave a more central role to some symbolic practices, notably Fourier series. The central move of the reorganization of his theory was the introduction of the hypothesis of natural radiation as a way to draw a divide between the macroscopic and microscopic state. Planck obtained his argument for irreversibility, but he had to pay a prize for it: his entire program depended essentially on the validity of Wien's law.

Keywords Resonator · Entropy · Irreversibility · Statistics · Boltzmann · Natural radiation · Molecular chaos

3.1 The Planck-Hertz Oscillator

Planck started his study of radiation theory well before Wien's derivation of the radiation law. In a famous letter to Robert Wood, he recalled that he had been struggling with the problem of equilibrium between matter and radiation since 1894 (Planck to Robert Wood, 7 October 1931). Indeed, this late piece of evidence matches early documents. In his *Antrittsrede* to the Prussian Academy of Science on 28 June 1894, Planck wrote:

© The Author(s) 2015
M. Badino, *The Bumpy Road*, SpringerBriefs in History
of Science and Technology, DOI 10.1007/978-3-319-20031-6_3

Likewise it is to be hoped that we shall be able to arrive at a better understanding also of those electrodynamic processes that are directly ruled by temperature, such as those occurring particularly in heat radiation, without being obliged to adopt the tedious roundabout way through the mechanical interpretation of energy. (Planck 1894, 1958, Vol. III, p. 3)

Thus, Planck's interest in heat radiation predates Wien's law and, I argue, was largely independent of the problem of deriving a radiation law for the black-body. The bulk of Planck's theory of radiation is contained in a series of five communications to the Prussian Academy of Science published between 1897 and 1899. I refer to this series as 'the Pentalogy'. Before starting the discussion on the Pentalogy, it is necessary to understand the main theoretical tool Planck deploys in his theory: the electromagnetic resonator.

The oscillating dipole introduced by Heinrich Hertz at the end of 1880s took by surprise almost all specialists who believed that electromagnetic waves were not produceable in a laboratory (Hertz 1889).[1] Although as an experimental object the dipole was a simple metallic artifact able to absorb and emit spherical electromagnetic waves, as a theoretical object, it was something much more mysterious. In his paper, Hertz carefully restrained himself from any specific assumption on the nature of the dipole.[2] Instead, Hertz first introduced a particular function of time and space that propagates as a spherical wave,[3] and only after showing that this potential provided a solution for Maxwell's equations, he proved that the field near the center of this spherical wave corresponded to the field generated by a dipole or a current oscillating harmonically.

In this way, the dipole as a theoretical tool acquired a particular flexibility. It could be used to investigate the features of a field in an empty space without introducing a source with special physical features. In fact, the ambivalence between oscillating dipole and oscillating current was maintained also in the first experimental investigations (Bjerknes 1891a, b). These characteristics of the Hertzian oscillator immediately attracted Planck's attention, and he devoted two important papers to its analysis. In the first paper, he investigates the conditions under which a dipole oscillating with a characteristic frequency and interacting with the field reaches a stationary state, i.e., the oscillation is sustained only by the energy absorbed by the field (Planck 1896, 1958, Vol. I, pp. 445–458).[4]

To determine the condition of stationarity, Planck uses an argument based on energy conservation. First, he considers a spherical region surrounding the oscillator and calculates the balance between incoming and outgoing energy. If the field is stationary, the net outgoing flux must be zero. To calculate the energy, he combines the primary field external to the oscillator with the field originated by the oscillator

---

[1] Important theoretical research on the propagation of spherical waves had been already carried out independently by Rowland (1884). For this reason, it is not uncommon to find in the literature the term 'Rowland-Hertz oscillator'.

[2] On this point see (Buchwald 1998).

[3] This function, which is actually a potential, is currently known as the Hertz vector. From this vector it is easy to derive the usual scalar and vector potentials (Bremmer 1958; Essex 1977).

[4] Planck assumes that no Joule effect takes place, and hence the energy emitted is due to the oscillation and not to transformation into other kinds of energy.

itself and applies Poynting's theorem. This gives three terms corresponding to the energy flux of the primary and oscillator fields plus the field due to their superposition. The first term can be dismissed: the source of the primary field is supposed to be outside the region, hence the case where only its field is present is uninteresting. The energy emitted by a free oscillation of the dipole had already been calculated by Hertz in 1889. The computation of the combined fields, instead, required some additional care.

Planck assumes that the primary radiation remains finite and continuous at any point in space except, obviously, at the source of the field. This allows him to expand in Taylor series the primary components at the center of the spherical region where the oscillator lies. We will meet this assumption of finiteness and continuity again in a completely different section of Planck's theory. As already pointed out by Hertz, in its immediate neighborhood (near field), the oscillator interacts only with the electric component of the field along its direction of vibrations, with all other components having zero effect on its motion. As a result, the net flux depends on the amplitude of the oscillation and on the phase difference between the oscillator and the exciting field. Therefore, Planck concludes, the oscillator is actually a resonator: it interacts with those field components with the same frequency and reaches the maximum when the phase difference is very small or zero (resonance).

In the last part of the paper, Planck surveys briefly the results of his analysis. Firstly, he insists that his approach holds only in thermal equilibrium: the resonant interaction, he states, allows one to study "stationary radiation processes occurring in a mechanical medium at rest, which is at a uniformly constant temperature and surrounded by bodies at the same temperature" (Planck 1896, p. 14). As a consequence, the resonator does not affect the energy distribution of the system, but it merely diffuses the radiation along all directions. This is a point I will return to later in the discussion of Planck's theory of radiation. Secondly, and related, the diffusion administrated by the resonator yields Kirchhoff's law of proportionality between emission and absorption but, and this is the point, not the specific form of the universal function.

In the second paper, Planck further generalizes his analysis (Planck 1897a, 1958, Vol. I, pp. 466–488). Instead of beginning with a harmonic solution of Maxwell's equations, he starts with a very general spherical wave and then cleverly exploits a condition on the linear size of the resonator to specify the properties of the field.[5] More importantly, here Planck applies his energetic argument to calculate the general equation of the resonator. According to the conservation principle, the time variation of the resonator energy must be equal to the outgoing flux calculated previously. Now, however, the calculation of the Poynting vector is complicated by the fact that

---

[5]In particular, he shows that if the resonator is small with regard to the wavelength of the interacting radiation, it is possible to simplify the expression for the dipole moment and its derivative (Darrigol 1992). Ultimately, Planck's theory of resonator relies on a clear-cut distinction between two lengths of different orders of magnitude: the characteristic wavelength involved in the resonator-field interaction and the geometric lengths ascribed to the material system (size of the cavity, linear dimension of the resonator). The latter must be much larger than the former to apply the usual laws of refraction and reflection without the intervention of diffraction phenomena.

Planck does not adopt an harmonic solution to begin with. The final result is a second order differential equation[6]:

$$L\ddot{f} + \frac{2K}{3c^3L}\dot{f} + Kf = Z_0, \tag{3.1}$$

where f is the dipole moment, Z_0 the component of the primary field driving the resonator, c the velocity of light, and K, L are characteristic constants of the resonator. This equation represents a damped oscillator driven by the external force Z_0. The damping constant can be cast in terms of the constants K, L or of the characteristic wavelength of the resonator:

$$\sigma = \frac{2\pi}{3c^3}\sqrt{\frac{K}{L^3}}. \tag{3.2}$$

Because the Joule effect had been excluded from the outset, this damping is solely due to the emission of energy: it is a radiative damping. But in a stationary process, the energy emitted, which is responsible for the damping, only comes from the primary field. This is exactly Planck's point: "by far the most important property of the radiative damping is that it fulfills the principle of conservation of energy without being necessary to add energy from outside" (Planck 1897a, p. 580). The conservative aspect of this damping differentiates it neatly from the usual frictional damping. Essentially, the radiation emitted amounts to a re-location of the energy previously stored in the primary field. Here, Planck adds an important comment: "the study of the conservative damping seems to me [...] of the utmost importance because by its means an outlook is open on the possibility of a general explanation of irreversible processes through conservative effects—a problem that the research in theoretical physics is presently tackling pushingly."

Intuitively, Planck's idea of explaining irreversibility through the damped resonator consists of two parts. Firstly, the diffusion of energy is likely to reach a stationary state without the addition of external work: it is a spontaneous process. Secondly, the interaction between resonator and field should be such that a reverse process is excluded, so that the stationary state remains in place forever. At this stage, nothing specific is said about how these two parts can be physically and mathematically substantiated. What is clear, however, is the role of the resonator to investigate the behavior of the radiation field. In Hertz's work, the resonator is an object; it is endowed with materiality. Planck stripped all materiality away and reduced it to an organized representation of a set of symbolic practices to manipulate the electromagnetic field. The resonator is symbolically codified by a mass, a natural frequency, two characteristic constants, and a dipole moment. This codification is then transformed by means of symbolic practices associated with Maxwell's electromagnetic theory and the mechanics of oscillating bodies. Planck turns the resonator in an alignment of symbolic practices: in his hands it yields a combination of thermodynamic-like

---

[6]Planck arrives at the second order from the usual third order equation by considering only real solutions.

arguments based on energy conservation and electromagnetic techniques and provides a probe with which to understand the behavior of the field without being wedded to questions on the specific microstructure of the sources. When forced to introduce the analyzing resonator, much later in his theory, Planck would follow exactly the same strategy.

3.2 The Pentalogy

3.2.1 The First Phase of Planck's Program

On 4 February 1897, Planck read at the Prussian Academy of Science a short communication on the irreversible processes of radiation (Planck 1897b, 1958, Vol. I, pp. 493–504). The obsession for irreversibility had always been the prime guideline of his work. Planck's doctoral dissertation, completed in 1879, had been devoted to the concept of entropy. The main claim was that the second law represents an absolute principle of nature, and entropy should not be limited to the equilibrium state because it expresses the degree of preference of nature for the stationary state (Planck 1879, 1958, Vol. I, pp. 1–61). During the 1880s, Planck's work on physical chemistry was strongly inspired by his conception of entropy and the second law. At the same time, he fiercely fought against the proponents of kinetic theory. Planck's dislike of kinetic theory had two main reasons. Firstly, although he was not in principle against the atomistic hypothesis, he felt that an interpretation of thermodynamics in terms of mechanical motion of microscopic particles cost a great deal of mathematical effort, which was disappointingly little rewarded (Planck 1891, 1958, Vol. I, pp. 372–381).

Secondly, in kinetic theory, the atomistic hypothesis goes hand in hand with the application of probabilistic methods, and those, in turn, result in a statistical interpretation of the second law. I discuss in depth this interpretation later on (see Sect. 3.2.2), though for the moment it suffices to say that, because in kinetic theory the equilibrium state was regarded as the most probable one, the statement that a thermal system tends to equilibrium admitted (probabilistic) exceptions. In fact, according to kinetic theory, the second law of thermodynamics states that the system tends to evolve from states of lower to states of higher probability, so the entire thermodynamics is a probabilistic business. Although Planck was not altogether contrary to the use of probability in physics, he could not agree with such a conclusion. In a very revealing letter to his friend Leo Graetz, he wrote:

> Probability calculus can serve if nothing is known in advance, to determine the most probable state. But it cannot serve, if an improbable state is given, to compute the following [state]. That is determined not by probability, but by mechanics. To maintain that change in nature always proceeds from [states of] lower to higher probability would be totally without foundation. (Planck to Leo Graetz, 23 May 1897)

The point of Planck's criticism is that probability provides us with useful rules for manipulating insufficient information, but it cannot replace the inner dynamics of the systems. Therefore, probabilistic techniques apply to physically accessible—albeit partial—information, but we still need to find explanations in terms of the dynamical evolution. However, in spite of Planck's opposition, at the end of the nineteenth century, kinetic theory was the only systematic attempt to understand the nature of irreversibility. Planck's fundamental project was to bring forward an alternative approach.

His first paper on radiation theory is divided into two parts. In the first, he describes the guiding principles of his work. The radiation contained in a cavity of reflecting walls is described by Maxwell's equations: its state does not change over time because it is a periodic, undisturbed motion. But if we introduce into the cavity a foreign element, it is possible to set up an irreversible modification in the system. Of course, Planck is referring to his resonator, which is able to alter the internal state of the cavity through the interaction with the field. In particular, the changes induced by the resonator in the surrounding electromagnetic field "have [. . .] a certain one-sided, equilibrating tendency" (Planck 1958, Vol. I, p. 495). The equilibrating effect of the resonator had already been partially described in the previous paper: basically it spreads the primary field all over the cavity. More specifically, the field becomes increasingly uniform because the fluctuations in the energy intensity are eliminated by the action of the resonator. This is the essence of Planck's irreversible process: the resonator irons out the irregularities in the space-time distribution of the field intensity. At the end of the process, the field is spatially uniform and it will not spontaneously return to a non-uniform state. There are three important points to note about this program.

First, the difference with Wien's approach is patent. Planck investigates the inter-action between the field and "something oscillating" with certain properties, but he does not touch upon the realizability of this interaction model. He does not question whether the resonator can really be a representation of a molecule, if there are conditions on the interaction with the radiation, and so forth. All the problems that had burdened Wien's and others' theories are here simply sidestepped. The representational aspect of the resonator almost coincides with its transformational properties: Planck's and Wien's programs differ fundamentally on their explanatory dimensions. To Planck, the generality of heat radiation is a justification for eschewing any microscopic assumption on the interaction mechanism.

Second, Planck's irreversible process consists in the redistribution of the energy intensity in space, which is a process that occurs separately for each resonator. Planck insists that the effect of the resonator amounts to a reduction of the space-time differences in the energy intensity, but on the energy distribution he has only a hazy comment to offer: "the resonator also influences the color of the exciting wave, if with this term one means the distribution of the total intensity on the different simple periodic vibration contained in it, and one may expect that also in this case the resonator acts in direction of a certain equilibration among the intensities of the different colors, whereof one can obtain important conclusions on the energy distribution of a stationary radiation state in the cavity considered" (Planck 1958, Vol. I, p. 496). This peculiar selectiveness of the resonator was totally at odds with contemporary research on heat radiation as Wien would hold much later (Wien 1909,

p. 302). More importantly, there was no concrete picture of how the resonators could mathematically, if not physically, affect the energy distribution.

Third, it is important to look at the conceptual resources Planck has at his disposal and the way he organizes them. Wien's main goal was the derivation of the radiation law. Because his program focused on a specific problem, he used thermodynamics and electromagnetic theory to narrow down the spectrum of possible solutions. To begin with, he derived general thermodynamic constraints on the law and then specified the form by kinetic and electromagnetic assumptions. Planck's program does not aim at the radiation law, but at irreversibility.[7] For this reason, although he relies on the same areas of knowledge, he arranges the representational and transformational dimensions differently. Strict thermodynamic irreversibility is the epistemic trope that shapes his epistemic story and affects the organization of his conceptual and technical resources. The goal is to describe electromagnetic processes that have the same character of irreversibility as the thermodynamic ones. Thus, the main difficulty sounds: how is it possible to integrate a fundamentally macroscopic body of knowledge (thermodynamics) with a microscopic one (Maxwell's electromagnetism)? How to set the micro-macro divide quickly became the driving theme of Planck's theory. The solution to this difficulty has two parts. The first, already partially accomplished, is to develop ideal objects, such as the resonator, and special techniques for manipulating the radiation, which are flexible and general enough not to overburden the final result with specific hypotheses. The second part, which is the essence of the Pentalogy, is the elaboration of an argument to prove strict irreversibility. Thus, Planck was handling the problem of heat radiation very differently from Wien. The distinction between theoretical dimensions allows us to see that although they were referring to the same areas of physical knowledge and often used similar concepts, their research programs diverged remarkably.

In the second part of the paper, Planck sets out a preliminary exploration of the field-resonator system. The solution of the differential equation of the damped resonator is the sum of a particular integral and the general solution. A crucial move here is the use of the Fourier series to represent the driving electric component Z_0 in the Eq. (3.1):

$$Z_0 = \sum_{n=1}^{\infty} A_n \sin \frac{2\pi nt}{T} + B_n \cos \frac{2\pi nt}{T}, \qquad (3.3)$$

where T is the time during which we follow the interaction. I comment extensively on the use of the Fourier series later on (see Sect. 3.2.4). At this stage, it is important to notice one typical assumption in Planck's theory. The time T must be taken to be much larger than the natural period τ_0 of the resonator, otherwise the final result

---

[7] Here one can see at work an example of the difference between 'physics of problems' and 'physics of principle' recently outlined by Seth (2010).

would depend on the choice of T.[8] Exploiting the flexibility of the resonator, Planck chose T so that $T/\tau_0 = n_0$ is an integer and σn_0 is a large number (σ is the logarithmic damping constant). The interesting consequence of this assumption is that one can neglect all the harmonics smaller than n_0 and confine the analysis to the higher frequencies.[9]

For a preliminary evaluation of the response of the resonator to the field, Planck relies again on energy conservation. The time-variation of the energy contained in a spherical region surrounding the resonator amounts to the time derivative of the dipole moment. Replacing this derivative with the fundamental equation of the resonator, Planck arrives at an equation expressing the change in the Z-component induced by the action of the resonator. More interestingly, if one uses the Fourier series, it is possible to express this effect in terms of the amplitudes and phase constants of the oscillating dipole.

3.2.2 The Statistical View of Irreversibility

Soon, Planck's program had to face the fierce criticism of Boltzmann (1897a, 1909, Vol. III, pp. 615–617). To understand the rationale of Boltzmann's critique, we must delve in the origins of statistical mechanics.

As early as the mid-1800s, it became popular among physicists to represent the thermal behavior of gases as the result of collisions between atoms or molecules. The basic idea was that, by occasionally hitting each other, particles exchange kinetic energy that, macroscopically, is perceived as heat. Ultimately, heat is a form of (microscopic) motion. In 1867, Maxwell established convincingly that a certain distribution of energy—or, alternatively, of velocities—among molecules exists that is not changed, on average, by molecular collisions (Maxwell 1867). This Maxwell's distribution has therefore the character of stationarity typical of thermal equilibrium. The question remained open whether it was the only distribution with such a remarkable property. A positive answer to this question, at least for the monoatomic ideal gas, came in 1872 when, in an epoch-making paper, Boltzmann showed that, under suitable conditions, Maxwell's distribution is indeed the only equilibrium distribution (Boltzmann 1872, 1909, Vol. I, pp. 316–402). Boltzmann's argument will play an important role in Planck's theory, and thus it is worthwhile discussing it in some detail.

An ideal gas is made up of molecules that move with vectorial velocities $\mathbf{v}$, have kinetic energy $m\mathbf{v}^2$ (m being the molecular mass), and undergo only binary collisions.

---

[8]These kinds of assumptions, similarly to the approximations on the linear size of the resonator, are common in Planck's theory and, more generally, in theories concerned with drawing macroscopic consequences from a microscopic model.

[9]Note that this consequence also follows from Wien's thermodynamic argument that only higher frequencies can produce thermal effects (Wien 1893). Though, Planck does not quote Wien: he wants to make a general argument on the behavior of an oscillating structure separate from the nature of heat radiation.

The macroscopic properties of this gas, e.g., its temperature, viscosity, heat conductivity, and so forth, can be derived by the distribution function $f(\mathbf{v}, t)d^3\mathbf{v}$, which tells us how many molecules are contained in the velocity interval $(\mathbf{v}, \mathbf{v} + d^3\mathbf{v})$ at time t. Boltzmann calculates a differential equation for the time evolution of the function $f(\mathbf{v}, t)d^3\mathbf{v}$ by evaluating the collision process. Let us assume that two molecules with velocities $\mathbf{v}_1, \mathbf{v}_2$ collide, and after the collision, they have velocities $\mathbf{v}'_1, \mathbf{v}'_2$, respectively. Boltzmann proves that the time derivative of the distribution function is:

$$\frac{\partial f(\mathbf{v}, t)}{\partial t} = \int_0^\infty \int_0^{\mathbf{v}_1+\mathbf{v}_2} d^3\mathbf{v}_2 d^3\mathbf{v}'_1 \psi(\mathbf{v}_1, \mathbf{v}_2, \mathbf{v}'_1) \left[f(\mathbf{v}'_1, t)f(\mathbf{v}'_2, t) - f(\mathbf{v}_1, t)f(\mathbf{v}_2, t) \right],$$
(3.4)

where $\psi(\mathbf{v}_1, \mathbf{v}_2, \mathbf{v}'_1)$ is the differential cross section that collects the geometrical details of the collision. The fundamental assumption behind Eq. (3.4) is the so-called *Stosszahlansatz* (SZA): the probability of a collision between a molecule of velocity $\mathbf{v}_1$ and a molecule of velocity $\mathbf{v}_2$ depends only on how many molecules with velocities $\mathbf{v}_1$ and $\mathbf{v}_2$ are present in the gas. Formally, the probability is proportional to the product $f(\mathbf{v}_1, t)f(\mathbf{v}_2, t)$.[10] From a probabilistic point of view, this amounts to saying that a molecule having velocity $\mathbf{v}_1$ is an event independent of another molecule having velocity $\mathbf{v}_2$. Equation (3.4) is the famous Boltzmann equation. To show that it tends to Maxwell's distribution, Boltzmann introduced the functional $H(f, t) = \int dt f(\mathbf{v}, t) \log f(\mathbf{v}, t)$. The time variation of H is related to the Boltzmann equation. Calculating the variation for the four possible velocities involved and averaging them, he obtained:

$$\frac{\partial H}{\partial t} = \frac{1}{4} \int_0^\infty \int_0^{\mathbf{v}_1+\mathbf{v}_2} d^3\mathbf{v}_2 d^3\mathbf{v}'_1 \psi(\mathbf{v}_1, \mathbf{v}_2, \mathbf{v}'_1) \left[f'_1 f'_2 - f_1 f_2 \right] \log \frac{f'_1 f'_2}{f_1 f_2}. \quad (3.5)$$

The form of the integrand function is such that it can only be positive or zero, and it is zero when $f(\mathbf{v}, t)$ is Maxwell's distribution. This proposition, known as the H-theorem, is a kinetic equivalent of the second law of thermodynamics, which is the statement that heat always flows from a warmer body to a colder one until equilibrium is reached.

Boltzmann's use of probabilistic language, techniques, and concepts suggested that something essentially new was embodied in the H-theorem. Ever since its foundation, kinetic theory had made large use of probabilistic assumptions and methods. They were an indispensable help to handle the complex interactions between innumerable molecules. Even when kinetic conclusions displayed an air of determinism, they held only 'on average', i.e., under provisos such as 'after waiting long enough...' or 'given a container big enough' and the like. It was only a matter of time before

---

[10]The complete definition of the SZA is more complex; see (Ehrenfest and Ehrenfest 1911, pp. 4–13).

someone started to ask questions about what would happens if odd—albeit physically possible—molecular arrangements are considered. As early as 1871, Maxwell argued that an imaginary being able to interact with individual molecules could create processes inconsistent with the second law (Maxwell 1871, pp. 308–309). Such a being could sit on the partition between two halves of a container and select molecules so to move the more energetic ones on, say, the left half. In this way, heat would 'spontaneously' (i.e., without spending work) flow from the colder body to the warmer. Maxwell's Demon made vividly apparent that violations of the second law were mechanically possible. Even more importantly, it showed that from the point of view of kinetic theory, irreversibility was merely a matter of high probability: because the Demon does not violate any physical law, one could obtain the same effect by means of molecular collisions only. In 1876, Joseph Loschmidt elaborated further on this point in a form that later became canonical (Loschmidt 1876). Let us imagine a system that has just reached the equilibrium state, with all its molecules arranged according to Maxwell's distribution. If we suddenly reverse all directions of motion, leaving unchanged the remaining parameters, the system will still be described by Maxwell's distribution, but it will begin tracing back its own evolution until it has again reached the initial state of non-equilibrium from which it started. This reversibility argument shows that the kinetic equilibrium is not absolute because it is always possible to conceive a microscopic state that is compatible with the equilibrium distribution but will evolve away from it over time.

Boltzmann was well aware of these consequences, and in 1877 he condensed his point of view in the claim that equilibrium is overwhelmingly the most probable state among the possible ones.[11] I dwell on the details of Boltzmann's 1877 combinatorial argument later on (see Sect. 4.5.1). For our immediate purposes it is important to mention a second objection against the statistical view of irreversibility. In 1896, building on the work of Poincaré, Ernst Zermelo, a former doctoral student of Planck, argued that, because a gas is a mechanical system evolving freely in a confined space, after a certain—very long—time it will pass through previous states. More specifically, let us assume that at instant t_1 the system is at state S_1; another instant t_2 exists in the future, such that at t_2 the system will be as close as one wishes to S_1 (Zermelo 1896). For Zermelo, this implied that the very idea of interpreting irreversibility as a property of a mechanical system was fundamentally incorrect. It implied that, if entropy is a function of state, it cannot increase to a maximum and remain constant thereafter. We are free to assume that S_1 is a non-equilibrium state, in which case at some point entropy must necessarily decrease to reach it again. Boltzmann was not impressed by Zermelo's argument. Besides the astronomically high value of the separation time $\Delta t = t_2 - t_1$, he insisted that a recurrence of non-equilibrium states was perfectly compatible with the statistical understanding

---

[11]Boltzmann (1877, 1909, Vol. II, pp. 164–223). According to common wisdom, Boltzmann did not realize the probabilistic implications of his theory until Loschmidt's reversibility objection (Brown et al. 2009; Klein 1973; Kuhn 1978). However, this claim does not stand before a careful examination of Boltzmann's papers during the period of 1868–1877. I have argued elsewhere that Boltzmann was aware since the late 1860s that irreversibility was a matter of probability (Badino 2011).

of irreversibility. The necessity of exceptions Zermelo had discovered was indeed entailed by the fact that equilibrium is only statistically stable (Boltzmann 1896, 1909, Vol. III, pp. 567–575).

Thus, since the late 1860s, Boltzmann had been defending strenuously, and in his view successfully, a conception according to which macroscopically absolute irreversibility was mere appearance, while at a microscopic level it was simply a matter of high probability. Being the world authority in statistical mechanics and the most prestigious upholder of the statistical view, he felt an urge to respond to Planck's program of tracing back "the one-directional processes to conservative effects" (Planck 1958, Vol. I, p. 493). Such a project, fairly clearly, was flying in the face of his firmest beliefs.

3.2.3 Boltzmann and Planck: The Human Side

The dispute around irreversible radiation process was the climax of a period of occasional, but bitter clashes between Planck and Boltzmann. The first conflict occurred at the time of the publication of Kirchhoff's lectures on theoretical physics. Planck edited the volume on heat theory and Boltzmann accused him of having badly misunderstood the meaning of the SZA (Boltzmann 1894, 1909, Vol. III, pp. 528–531). In the derivation of Maxwell's distribution, Kirchhoff, or the editor of the volume according to Boltzmann, had assumed that the SZA held both for an arbitrary collision and for the reverse collision that is obtained by reversing the velocities of the molecules. Boltzmann contended that in the second case, the collision cannot be treated as a combination of independent events. Charged with professional misbehavior, Planck reacted bluntly and defended his interpretation of Kirchhoff's text (Planck 1895, 1958, Vol. I, pp. 442–444). He argued that, although Boltzmann was right in claiming that the SZA does not hold for the reverse collision, it was precisely a crucial feature of Maxwell's distribution that it was not changed by collisions occurring in the reversed direction. Planck's reply missed the point Boltzmann wanted to make, namely that some sort of 'chaos' assumption was implicit in the derivation of Maxwell's formula.[12] Therefore, before ending the dispute, Boltzmann stressed that for the proof of the uniqueness of Maxwell's distribution, "the assumption is necessary that the state of the gas is disordered and remains so, namely [...] the frequency of each kind of collision can be derived from the laws of probability" (Boltzmann 1895b, 1909, Vol. III, pp. 532–533).

The relationship between Boltzmann and Planck could not recover from this incident, which can be seen from an episode that occurred during Boltzmann's controversy with Zermelo. At that time, Planck was already one of the editors of *Annalen der Physik* and Boltzmann suspected that he might profit off this position

---

[12]Indeed, Planck was right: Maxwell's distribution is the only one that fulfills the SZA applied to reversed collisions (Ehrenfest and Ehrenfest 1911, pp. 11–13). In that case, however, the SZA ceases to be a probabilistic assumption.

by favoring his former student. For this reason, he wrote to Eilhard Wiedemann, son of Gustav Wiedemann, editor-in-chief of *Annalen*, to support his case. In the letter, he listed what, in his opinion, were minimal conditions of fairness: "I think I'm entitled to demand: (1) that Herr Planck does not delay the publication of my essay, (2) that not a word will be changed of it, (3) that a reply to it does not appear on the same issue."[13] As far as we can judge, Boltzmann's demands were satisfied, although, from the letter to Leo Graetz mentioned above, we know that Planck supported almost completely Zermelo's objection to Boltzmann's theory.

On the basis of this deep divergence of research style and even mutual personal diffidence, there is little wonder that Boltzmann immediately and sharply reacted to Planck's program. The gist of Boltzmann's objection is that the resonator cannot change irreversibly the cavity radiation because (1) it requires Maxwell's equations, which are as reversible as any good mechanical equation, and (2) a resonator works as a reflecting mirror, so it distributes but does not change the radiation density. To drive this point home, he proposes an analogy with kinetic theory. Let us imagine that the resonators are fixed obstacles placed randomly in the cavity and the radiation is a collimated bundle of particles. Elastic collisions against the fixed obstacle will make the motion more and more uniform in space, but nothing in this process is irreversible. Thus, Boltzmann concludes, "all one-sidenesses that Herr Planck finds in the action of the resonators, hinges on the choice of one-sided initial conditions" (Boltzmann 1909, Vol. III, p. 616). Furthermore, Boltzmann rebuts Planck's allusions on the inadequacy of kinetic theory in handling the issue of irreversibility. Kinetic theory, Boltzmann argues, resisted many attempts at demolishing it and met all the challenges. Instead of denigrating it, Planck would better follow its example and discover a mathematical expression playing "a role analogous to entropy." Unexpectedly, Planck would follow this advice in due course.

Planck replied some weeks later with the second installment of the Pentalogy (Planck 1897c, 1958, Vol. I, pp. 505–507). From this short note, we can get a fairly clear idea of Planck's argumentative pattern to counter the reversibility objection: the reverse process cannot be accepted because it does not meet some conditions of the problem. The rationale of Planck's idea is the notorious asymmetry between advanced and retarded solutions for the wave equation. As we have seen, the Hertz vector propagates as a spherical wave. D'Alembert's general solution of this equation consists in the superposition of two waves, one expanding from and one contracting on the source of the wave. There was, and sometimes still is, a heated debate on the physical admissibility of these two solutions. The most popular view was that the advanced solution (the inward one) must be rejected because (1) it is hard to figure out a physical source producing a contracting spherical wave, (2) it presupposes that future states of the wave act on past states, and (3) the energy intensity increases as the radius decreases and therefore it should be infinite at the source.

This third point was particularly useful to Planck. He had explicitly required finiteness and smoothness at the singular point as a condition for the primary wave. But an inward wave does not comply with this condition; hence, it cannot be an acceptable

---

[13]Boltzmann to Eilhard Wiedemann, 20 March 1896, (Höflechner 1994, Doc. 427).

exciting wave for Planck's resonator. The advanced part of the general solution of the wave equation as well as any inward wave must be discarded. Planck concludes his paper by returning the unfriendliness to Boltzmann: without mentioning explicitly kinetic theory, he states that his own approach "seems [...] to offer the perspective for a foundation of a rational theory of irreversible processes, contrary to that conception that misplaces the conditions of irreversibility in the initial conditions of the world" (Planck 1958, Vol. I, p. 507).

Boltzmann's further reply came on 18 November and was basically a reiteration of the previous arguments with some additional taunts (Boltzmann 1897b, 1909, Vol. III, pp. 618–621). He accused Planck of constructing a "straw man theory" in which an arbitrary assumption without any physical ground is introduced with the sole goal of excluding the reverse process from the outset. Nobody disputes that this move is always possible, Boltzmann says, but the ensuing theory would be a purely formal one. There is a subtext hidden in this statement. Boltzmann is implicitly referring to the debate on the H-theorem that raged on the columns of *Nature* in 1894–95 and that I discuss below (see Sect. 3.2.6). Ultimately, however, Boltzmann was challenging Planck to show a concrete analysis of the way in which radiation theory could avoid the reversibility argument. Furthermore, he also reminded Planck that, ironically, his radiation theory was as exposed to Zermelo's recurrence objection as kinetic theory. Planck decided that the best way to meet the challenge was to thoroughly pursue his argument for the irreversibility in the cavity radiation.

3.2.4 Planck's Third Paper and Boltzmann's Ultimate Criticism

It took some months for Planck to complete a full-fledged theory of cavity radiation. The third paper of the Pentalogy was read only at the beginning of 1898 (Planck 1898a, 1958, Vol. I, pp. 508–531). In this systematic treatment, Planck tackles the three parts of his explanatory dimension for irreversibility: (1) the elimination of reverse processes, (2) the analysis of the recurrence problem, and (3) the description of the final stationary state.

The problem looks very different with or without the resonator. In the first case, Planck lays down Maxwell's equations in the Hertz version:

$$\mathbf{E}_x = \frac{\partial^2 H}{\partial x \partial z} \qquad \mathbf{B}_x = \frac{1}{c}\frac{\partial^2 H}{\partial y \partial t} \qquad (3.6)$$

$$\mathbf{E}_y = \frac{\partial^2 H}{\partial y \partial z} \qquad \mathbf{B}_y = -\frac{1}{c}\frac{\partial^2 H}{\partial x \partial t} \qquad (3.7)$$

$$\mathbf{E}_z = \frac{\partial^2 H}{\partial z^2} - \frac{1}{c^2}\frac{\partial^2 H}{\partial t^2} \qquad \mathbf{B}_z = 0, \qquad (3.8)$$

where **E**, **B** stand for the electric and magnetic strengths, respectively, c is the velocity of light, and the Hertz vector H obeys the wave equation:

$$\frac{1}{c^2}\frac{d^2 H}{dt^2} = \Delta H. \tag{3.9}$$

The general solution of this equation is:

$$H = \frac{1}{r}\left[\phi_1\left(t - \frac{r}{c}\right) + \phi_2\left(t + \frac{r}{c}\right)\right], \tag{3.10}$$

where ϕ_1, ϕ_2 are the potentials representing an onward and inward wave, respectively. These are the basic tools to describe the behavior of the radiation in an empty cavity. To complete the electromagnetic problem, one has to add suitable boundary conditions. Both at the singularity point (the center of the cavity) and at the walls, the Hertz vector must be zero. These boundary conditions entail that the process is periodic: the radiation consists of a set of concentric wave fronts expanding and contracting regularly. Consequently, the Hertz vector can be represented by a Fourier series. Now, the energy carried by these waves during a time interval T can be calculated by applying Poynting's theorem. There is a technical problem, though. By squaring the Fourier series, one gets the simple squares corresponding to the energy individually carried by each harmonic and the cross products that originate interference terms. Macroscopically, we do not measure instantaneous energy, but rather average energy over a period of time long in comparison with the period of vibrations of the Fourier components.[14] The time average of the energy is called radiation intensity and can be cast in Fourier series as follows:

$$J = \frac{1}{2}\sum_n C_n^2 + \sum_a\left[A_a \sin\frac{2\pi a}{T}\left(t - \frac{r}{c}\right) + B_a \cos\frac{2\pi a}{T}\left(t - \frac{r}{c}\right)\right], \tag{3.11}$$

where the Fourier coefficients A_n, B_n are in turn harmonic functions[15]:

$$A_a = \sum_n C_{n+a}C_n \sin(\theta_{n+a} - \theta_n)$$

$$B_a = \sum_n C_{n+a}C_n \cos(\theta_{n+a} - \theta_n).$$

This series consists of two parts. The amplitude C_n^2 in the first term expresses the intensity associated with each individual harmonic and is time-independent. The

---

[14]That such a time exists, follows from the condition above that only the harmonics with very high frequency and short period play a role in heat radiation.

[15]The amplitude C_n is related to the amplitude of the Hertz vector, while θ_n are its phase constants. The index $a = 1, 2, 3, \ldots$ comes from an ingenious rearrangement of the harmonics that Planck operates to simplify the calculation of the Poynting vector. The cross products can be ordered in groups of harmonics separated by an increasing distance a.

second term is a combination of harmonic functions with amplitudes that are in turn functions of the time. It represents the fluctuations of intensity at different points of space and time as the result of the interference between the components: "the constant first term of the series designates simply the sum of the radiation intensities of all partial vibrations [...] the variability of the radiation intensity with time and place depends on the values of the coefficients A_a and B_a; only when each of them has a negligible value in comparison with the first term the radiation intensity is constant" (Planck 1958, Vol. I, p. 518). Here, we can see a clear example of epistemic cooperation: quantities C_n^2, A_a and B_a codify the intensity fluctuations only insofar as they are included in a mathematical technique to manipulate radiation intensity, i.e., Fourier series.

The introduction of a resonator in the center of the cavity substantially changes the boundary conditions. The resulting field is the superposition of the free radiation and the outward wave emitted by the resonator. At the singularity point, the field is given by the usual resonator equation driven by the Z-component of the electric field. At the walls, the periodicity of the process requires the sum of the radiation and resonator field. Looking for periodic solutions, Planck determines easily the form of the potential in the Hertz vector and of the electric moment of the dipole[16]:

$$\phi(t) = \sum_n D_n \cos\left(\frac{2\pi k_n t}{T} - \theta_n\right) \tag{3.12}$$

$$f(t) = -2\sum_n D_n \sin \pi k_n \sin\left(\frac{2\pi k_n t}{T} + \pi k_n - \theta_n\right). \tag{3.13}$$

Having laid down the formal machinery, Planck moves on to show why and in what sense this electromagnetic process is irreversible. The first problem to tackle is the reversibility objection: "a direct reversal of the process must be excluded" (Planck 1958, Vol. I, p. 509). Planck's strategy is to show that the process that is obtained by changing the variable t into $-t$ in the equations above is not compatible with the general conditions of the problem. According to Planck, the time reversal of the process given by Eqs. (3.12) and (3.13) is:

$$\phi_r(t) = -\sum_n D_n \cos\left(\frac{2\pi k_n t}{T} + \theta_n\right) \tag{3.14}$$

$$f_r(t) = 2\sum_n D_n \sin \pi k_n \sin\left(\frac{2\pi k_n t}{T} - \pi k_n + \theta_n\right). \tag{3.15}$$

It is very easy to show that these equations do not fulfill the boundary conditions; therefore, they are not acceptable solutions. It is important to stress this point: Planck's response to the reversibility objection amounts to proving that it is possible

---

[16]The integers k_n are related to the natural period of the resonator. Essentially, they select the acceptable solutions in terms of the resonance interaction. The amplitude D_n is related to the field amplitude C_n via special phase parameters.

to set up a complete electromagnetic problem (with suitable equations and boundary conditions) such that the time reversal of a solution is not, in turn, a solution. In other words, the irreversibility is inherent in the formal features of the problem. This is Planck's *first argument for irreversibility*.

To the issue of recurrence, Planck has a much more complex solution. He concedes that the mechanic-like character of the problem leaves open the theoretical possibility of a reappearance of some previous states in accord with Poincaré's and Zermelo's arguments. However, this recurrence is constrained by special conditions. Planck introduces an important distinction between two classes of processes. On the one hand, there are processes in which the radiation intensity of the primary field is not uniformly distributed over the Fourier components, but a few of them contain almost all the total intensity. Alternatively, the intensity may be distributed over many monochromatic components, but they in turn are distributed regularly in the radiation: for example, only every other harmonic carries an intensity sensibly different from zero. In this case, the process is referred to as 'tuned to the system' because it is clear that such a primary field excites the reaction of a specific class of resonators only.

On the other hand, if none of the previous conditions materialize, the process is 'not-tuned to the system.' This means that the intensity allocated over each component is very small in comparison with the total intensity, and there are no regular gaps in the distribution of the intensity. By a simple argument, Planck shows that the recurrence time is tightly related to the structure of the radiation. In particular, this time increases enormously if there are many 'active' components in the primary field and becomes much longer than any experimentally conceivable time. Thus, Planck concludes, "if [. . .] the radiation is 'not tuned to the system', i.e., if the partial vibrations are disposed either seamlessly or with irregular gaps, then for a finite, not at all small, number of them a recurrence of a previous state is excluded for all times in which [our] equations allow a conclusion" (Planck 1958, Vol. I, p. 529).

These assumptions on the disposition of the Fourier components in the primary field become much less surprising when we take a closer look at the technique used by Planck to handle the radiation. As we have seen, Planck consistently deployed the Fourier series to represent the measurable quantities of his theory, electric components, dipole moment, intensity, and so on. This approach was not as common as we may think at first sight. Wien, for one, never used the Fourier series, and we will see in a moment why. In using this particular technique, Planck was placing himself in a different theoretical tradition. Historically, the fact that white light, light made up of many frequencies, does not display interferences between its own components in experiments with gratings was seen as an argument to conclude that the components were disposed regularly. In 1889, Louis Georges Gouy (1854–1926) suggested that this conclusion was a bit hasty because the regularity could have been originated by the grating itself. In reality, Gouy continued, white light may be made up of irregularly distributed impulses that can be represented as a whole in a Fourier series (Gouy 1886). This idea was taken up by Lord Rayleigh (1842–1919), who built on an analogy with acoustic to stress two points: (1) contrary to the acoustics case, light vibrations were "inextricably blended," and (2) radiation represented by experimental laws such as Weber's can be interpreted as originating from irregular harmonic pulses distributed according to a law of error (Rayleigh 1889, pp. 462–464). This way of handling the radiation was further developed in the last decade of the nineteenth century before being taken up by Planck (Schuster 1894; Wien 1909, pp. 344–345). In this view, thus, the single radiation components do not have individual existence and cannot be experimentally isolated. This picture was miles away from Wien's (and others') idea that radiation is constituted by independent monochromatic components that can be dealt with as material molecules.

But the larger point concerns the epistemic surplus encapsulated in the formal technique of the Fourier series. For Planck, this mathematic practice conveys a general picture of the field in which macroscopic quantities are made up of components ordered in very complex arrangements. Using Fourier series to handle field quantities amounts to treating these quantities as constituted of microscopic vibrations that have no independent physical meaning. This representation differs remarkably from the one deriving from usual mathematical methods of kinetic theory. The conceptual consequences of this difference will be fully appreciated when I discuss the hypothesis of natural radiation (see Sect. 3.2.5). Here, it suffices to note that the representation of the field embodied in the Fourier series also constitutes a common ground for analogies and conceptual exchanges between radiation theory, acoustics, and even statistical mechanics. Within this context, it is not surprising that Planck made use of disorder assumptions: some sort of hypothesis on the disposition of the components was already implicit in the application of the Fourier formalism to radiation.

Finally, if the process is unidirectional, there must be a stationary state. Intuitively, Planck imagines that the field-resonator interaction produces a secondary wave in which the fluctuation terms are smaller than in the primary wave. After a sufficient number of interactions, the fluctuations become negligibly small, and the radiation intensity maintains a constant value throughout the cavity. This is the final stationary state. Planck, however, does not put forward a mathematical description of this process. At this point, it is still a rough idea. Instead, he does discuss the general conditions for the stationary state. The fluctuation terms of the Eq. (3.11) are series of trigonometric functions; therefore, they will die out if their phase constants do not produce constructive interferences between neighboring terms: "for a measurable fluctuation of the total radiation in a non-tuned wave it is necessary an appreciable interference between neighbor partial vibrations and a certain law-like regularity [Gesetzmässigkeit] in the values of the phases: the process must be 'ordered' in a certain sense" (Planck 1958, Vol. I, p. 531). A stationary state is therefore a perfectly disordered one in which no special arrangement of the components produces constructive interferences.[17] This closes Planck's third paper.

---

[17] A clarification of this argument would come in the fourth paper. There, Planck shows that the Fourier series of the radiation intensity is time-dependent in two different ways: in the main series, the time appears in periodic terms, while in the Fourier coefficients, time is the argument of functions that are in general aperiodic. The equilibration process acts precisely on these aperiodic terms (Planck 1898b, 1958, Vol. I, p. 541).

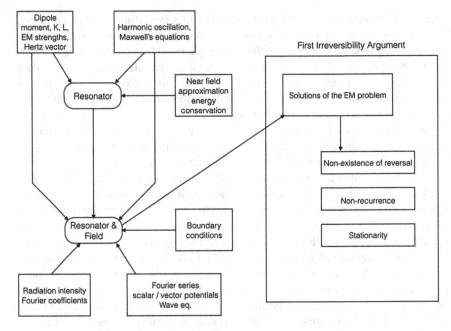

Fig. 3.1 Planck's theory in the third part of the Pentalogy (1898)

Let us pause a moment to examine the morphology of this theory (Fig. 3.1). One can see that there is no clear center from which the rest of the theory derives. Instead, Planck combines the representational dimension (notions such as the dipole moment, the electric and magnetic strength, and radiation intensity) with a set of tools to transform symbolic representations (Maxwell's equations, Fourier series, approximations, and boundary conditions). Cooperations between dimensions occur at different stages of the theory to solve specific problems. The first part is the description of the resonator and the derivation of its equations. Next, Planck discusses the interaction between resonator and field. This analysis requires some symbolic codifications and practices already introduced, as well as new ones. At this second stage, Planck seeks for the solution of the electromagnetic problem in terms of potential ϕ and dipole moment f. This solution is then the basis for the first argument of irreversibility, which in turn uses some of the codifications and practices of the other stages. The entire theory is aligned along a pretty linear story: the field-resonator interaction is an irreversible process because the equations representing it cannot be time-reversed.

The clarity of the argument for irreversibility allowed Boltzmann to strike back with a definitive blow. From his correspondence, we know some details about the final part of the controversy. In a letter to his friend Felix Klein, Boltzmann mentions Planck's recent paper:

Perhaps you have quickly glimpsed at the dispute I have pursued with Herr Planck in the Berlin Academy under the heading 'irreversible processes.' Now in a communication of the 16 December, Herr Planck has reversed the exciting wave for a specific case, but he has completely forgotten that the wave prior emitted by the resonator must be reversed as well. From the circumstances that he has obtained a totally counterintuitive formula, he has not concluded that he was wrong, but rather that he had found out a process whose reversal is not possible. I have sent him directly my considerations, a move that will not necessarily shorten the dispute; I'm curious to hearing his response. (Boltzmann to Felix Klein, 12 February 1898, (Höflechner 1994, Doc. 462))

Shortly after Boltzmann returns on the issue with what sound like closing words from his quarter:

As for the controversy I had with Planck in the Berlin Academy, I wrote him still a postcard not to continue it in this form. The reason is that Planck wrote me that he acknowledges that I'm right, that he himself will present my essay to the Berlin Academy and that he is projecting a new essay in which, although with reservations, he will accept my objection. Thus, the thing has been settled over, at least temporarily; for I don't have the slightest idea about what Planck's expected paper will look like. The story has excited me a little. I could not understand that Planck did not grasp my arguments and I believed, said between you and me, that he did not want to admit his mistake and was hedging around. As far as sincerity and honesty are concerned, I estimate him much more now. (Boltzmann to Felix Klein, 15 March 1898, (Höflechner 1994, Doc. 465))

What was this ultimate objection about? As touched on in the letter of 12 February, Boltzmann argues that what Planck has presented as the time reversal of the electromagnetic process is, in point of fact, a gross mistake (Boltzmann 1898b, 1909, Vol. III, pp. 622–628). Briefly stated, in construing his time reversal, Planck forgot to reverse the Hertz vector as well. To make this point as clear as possible, Boltzmann derives a general functional relation between the original and the reversed potentials for Planck's specific case. Then he proves that Planck's equations (3.14) and (3.15) do not fulfill the condition of this functional, and therefore they cannot possibly be the correct reversal. To get the right solution, one has to modify the potential as follows:

$$\phi_r(t) = -\sum_n D_n \cos\left(\frac{2\pi k_n t}{T} + \theta_n\right) - 2\sum_n D_n \sin \pi k_n \sin\left(\frac{2\pi k_n t}{T} - \pi k_n + \theta_n\right).$$

$$(3.16)$$

It is easy to show that this potential, together with the dipole moment of Eq. (3.15), fulfill Boltzmann's condition for the reversal, as well as Planck's boundary conditions. Hence, Boltzmann's argument amounts to proving that S is a solution of Planck's problem *if and only if* its time reversal is a solution. The morphology of Fig. 3.1 is totally disrupted because the argument for irreversibility no longer works. In the last part of the note, Boltzmann explains the mistake. The electromagnetic process consists of the superposition of the primary wave and the wave given off by the resonator. Planck's calculation of the reverse process simply reversed the wave produced by the sole resonator. In reality, the reverse of the primary wave is the superposition of two reversals: the resonator wave and the radiation wave.

This argument by Boltzmann did not leave any way out: Planck had to give up completely his original idea of proving irreversibility as the consequence of a well-defined electromagnetic problem. Planck's ambitious program had not stood the most classical of the objections to irreversibility, and a complete reorganization was apparently necessary.

3.2.5 Facing Failure: The Reorganization of Planck's Program

Planck's initial hope was to prove irreversibility as the straightforward consequence of a well-conceived electromagnetic problem. He succeeded in restricting the effects of recurrence and had a fairly clear idea of the stationary state he wanted to reach, but he failed woefully in dealing with the issue of irreversibility, as Boltzmann was ready to stress. However, Boltzmann did not leave only destruction behind him. His deep criticisms suggested a possible alternative: to follow a strategy inspired by kinetic theory. But Planck had to be very careful in picking up this option. There was probably something in kinetic theory that could be helpful in fathoming the irreversibility riddle, but, at the same time, Planck remained convinced that there was also something profoundly wrong with that approach. While Wien had happily deployed kinetic arguments and had even conceived of an electromagnetic analog of Maxwell's Demon, the epistemic trope of Planck's theory still pointed at a story of the second law that was very different from kinetic theory. What Boltzmann convincingly showed by means of his criticism was that this story could not possibly rely solely on the general features of the electromagnetic problem. Therefore, Planck had to find a way to realize the explanatory dimension of his theory by means of a new argumentative pattern.

These considerations led to a reorganization of Planck's program along two main guidelines. First of all, he incorporated the techniques and argumentative pattern of the H-theorem. This argumentative pattern consists of three steps: (1) introducing an entropy function, (2) calculating its time evolution, and (3) proving that this time evolution monotonically reaches a stationary state. The problem, however, was that this argument was traditionally carried out by means of mathematical practices related to probability theory. Planck's attitude toward probabilistic methods has been well described in his letter to Leo Graetz: they concern the manipulation of physically accessible information in a regime of epistemic uncertainty. The deep reason of the statistical view of irreversibility lies in the epistemic surplus that Boltzmann reads off from his methods. He claims that they concern the microscopic constituents, i.e., irreversibility is inherently statistical. But Planck thinks differently. Probabilistic methods and assumptions do not entail anything beyond our epistemic incapacity of interacting with the ultimate furniture of the universe.

Planck reads a different epistemic surplus in probabilistic practices, and to substantiate it he must introduce a special hypothesis to create a divide between the accessible macroscopic world and the inaccessible microworld of the radiation cavity. This is the second guideline of his program. To reconfigure the traditional epistemic surplus of kinetic arguments according to his view, Planck must black-box the microscopic constituents of radiation and concentrate the argument on the manipulation of macroscopic quantities only. This line of reasoning resonates perfectly with his predilection for pure thermodynamics. Whereas in the third paper Planck insists on the fact that the field-resonator interaction depends on microparameters such as amplitudes and phase constants, from the fourth paper onwards he focuses on a description based on averages and macroscopic quantities. However, it is not enough to say that microscopic quantities can be ignored: it must be shown how they can be substituted by macroscopic parameters. For this, Planck introduces the hypothesis of natural radiation. This hypothesis serves the reconfiguration of the symbolic codification of the theory, is grounded in the use of the Fourier series, and prepares the explanation of irreversibility: it is the result of a true epistemic cooperation between the various dimensions of Planck's theory.

Planck worked on the wholesale reorganization of his program throughout the first half of 1898 and finally produced the fourth chapter of the Pentalogy on 14 July (Planck 1898b, 1958, Vol. I, pp. 532–559). This paper is much longer than the previous ones. The numbering of sections and equations continues from the third installment in an attempt to show continuity. His first step is the elaboration of two concepts: the radiation intensity and the average energy of the resonator. While previously Planck had treated the electromagnetic process using mainly the Hertz vector, here he replaces the microscopic potentials with the radiation intensity, that is, the time average of the energy. Of course this quantity satisfies the same wave equation and the same boundary conditions. We can therefore distinguish between a primary intensity $J(t)$ due to the exciting wave and a secondary intensity $J'(t)$ due to the superposition between the primary wave and the wave emitted by the resonator. The radiation intensity is a macroscopically measurable energy, and the difference between the secondary and the primary intensity gives the time variation of the average energy of the resonator.[18]

After casting the intensities in Fourier series, Planck moves on to average the resonator energy and to express it again in a suitable series[19]:

$$U = \frac{b_0}{2} + \sum_a a_a \sin \frac{2\pi a t}{T} + b_a \cos \frac{2\pi a t}{T} \tag{3.17}$$

---

[18]In fact, the energetic difference between the primary and the secondary field is precisely the contribution of the resonator.

[19]The parameter γ confines the series to terms very close to the characteristic frequency of the resonator.

$$b_0 = T \sum_n C_n^2 \gamma$$

$$a_a = T \sum_n C_{n+a} C_n \gamma \sin\left(\frac{2\pi a \gamma t}{T} + \theta_{n+a} - \theta_n + \pi a \gamma\right)$$

$$b_a = T \sum_n C_{n+a} C_n \gamma \cos\left(\frac{2\pi a \gamma t}{T} + \theta_{n+a} - \theta_n + \pi a \gamma\right).$$

In this way, Planck can treat the electromagnetic process by means of macroscopic quantities only. Because the time variation of the resonator energy is $\frac{dU}{dt} = J(t) - J'(t)$, the process boils down to a relation between the corresponding Fourier coefficients. This is the gist of Planck's new approach.

Between radiation intensity and resonator energy, an electromagnetic interaction takes place, which, in the previous paper, was described in terms of potentials and dipole moment. Here, Planck changes tack and studies the way in which the radiation intensity is modified by the resonator energy. However, the intensity of the primary radiation is made up of many components, while the resonator interacts only with the component with the same characteristic frequency. It is necessary to decompose the primary intensity into its spectral components at different frequencies. Here comes the crucial move. The Fourier components represent a ready-made spectral decomposition of the intensity, and they are extensively used in this sense in acoustics. More importantly, in the conclusions of the third paper, Planck had somewhat suggested that the resonator interacts with the Fourier component at the resonance frequency (Planck 1958, Vol. I, pp. 530–531). Here, however, he develops another line of reasoning.

The spectral component of the intensity must have a *physical meaning*, whereas "a single member of the Fourier series has absolutely no independent physical meaning as long as it cannot be physically isolated and measured" (Planck 1958, Vol. I, p. 534). It is impossible to conceive an analyzing device that, like the tuning fork, is tuned on a single Fourier component. To support this point, Planck resorts precisely to the theoretical tradition within which he had treated the radiation. He mentions Gouy's experiments to conclude that "the fact that one has not succeeded in producing an interference between two rays of the same color stemming from different sources is equivalent to the fact that it is impossible to produce absolutely homogeneous light" (Planck 1958, Vol. I, p. 535). From the impossibility of technically producing monochromatic light, Planck infers the physical meaninglessness of the individual Fourier component.

A point is worth stressing here. Wien treated the micro-macro distinction in radiation theory in close analogy with kinetic theory, but that strategy led to Maxwell's Demon and the statistical view. Planck has to look for a different way. Macroscopic quantities, such as radiation intensity, are made up of microscopic field components, amplitudes, and phase constants, but those quantities have no physical meaning individually. Therefore, they were not suitable for formal manipulation as in kinetic theory. Physics deals with macroscopic quantities and not with the multitude of field components constituting them. In making this move, Planck is using the epistemic surplus of the Fourier series. The tradition of Gouy and Rayleigh, to which Planck was appealing, considered the Fourier series as a useful formal tool to handle radiation. Planck's discussion on the spectral decomposition, however, singled out the Fourier series as a way to display formally the relation between microscopic and macroscopic description. The Fourier series comes to represent a collective effect of many uncontrollable microscopic entities, a symbolic role somewhat analogous to that played by the distribution function in kinetic theory. Unlike kinetic theory, however, the microscopic constituents are deprived of any individual physical meaning. Amplitudes and phases make physical sense only within the Fourier representation. This passage of Planck's theory shows how the transformational dimension can cooperate with the representational one. Planck has a concrete problem: how to handle the field-resonator interaction in terms of macroscopic (time-averaged) quantities. The Fourier series are not simply a way to transform a symbolic codification into another, but they provide a general information about the process, i.e., that macroscopic interaction is the result of innumerable microscopic amplitudes and phases.

The conclusion of this argument is that a physically meaningful spectral component of the intensity is not an individual Fourier component, but rather it is "represented by a large number of neighbor partial vibrations" (Planck 1958, Vol. I, p. 534). While the individual components are microscopic objects completely outside our control, the spectral decomposition must still be a macroscopic entity. Planck is aware that there was "no attempt at providing a mathematical definition of the radiation intensity at a single color" (Planck 1958, Vol. I, p. 533). The concept is indeed a byproduct of his strategy of confining the analysis of radiation within a macroscopic level. Interestingly, Planck justifies this move with the appeal to the 'physical meaningfulness'. This suggests that the concept of spectral decomposition occupies a central role in the reorganization of his program: it creates, within the strictures of Planck's program, the conceptual space to interpret the relation between macroscopic quantities and microscopic constituents consistently with his epistemic trope. To see how this happens, we have to delve further into the argument.

The spectral decomposition of the intensity can be obtained through a process of ideal measurement by means of an ideal resonator tuned on a certain frequency. The crucial parameter of this 'analyzing resonator', as Planck calls it, is the damping constant: it must be small enough to respond to an infinitesimal interval around the harmonic, but large enough to stop the vibration rapidly.[20] Planck defines the spectral component of intensity at frequency v as the energy emitted by the analyzing resonator tuned on the frequency v. Because the analyzing resonator reacts to the characteristic frequency, but also to frequencies infinitesimally close to it, the spectral component is still made up of many monochromatic components and can be cast in Fourier series[21]:

---

[20]If the damping process lasts too long, the resonator will still be in vibration when a new train of waves impinges on it, and the energy given off will be the superposition of the radiations that arrived at different times instead of the component of a single wave.

[21]The parameter δ plays the same role as the parameter γ: it constraints the series in order that only frequencies very close to the characteristic frequency carry a non-negligible energy.

$$J_v(t) = \frac{\mathbf{B}_0^v}{2} + \sum_a \mathbf{A}_a^v \sin \frac{2\pi at}{T} + \mathbf{B}_a^v \cos \frac{2\pi at}{T} \tag{3.18}$$

$$\mathbf{B}_0^v = T \sum_n C_n^2 \delta$$

$$\mathbf{A}_a^v = T \sum_n C_{n+a} C_n \delta \sin \left(\frac{2\pi a\gamma t}{T} + \theta_{n+a} - \theta_n \right)$$

$$\mathbf{B}_a^v = T \sum_n C_{n+a} C_n \delta \cos \left(\frac{2\pi a\gamma t}{T} + \theta_{n+a} - \theta_n \right).$$

Planck shows that if we integrate $J_v(t)$ over all frequencies we arrive at the total intensity $J(t)$. Furthermore, the spectral component $J_v(t)$ satisfies the boundary conditions of the problem, therefore it is a well-defined quantity. Unsurprisingly, it turns out that the spectral component of the intensity does not give the amplitudes C_n and the phase constants θ_n of the primary field: "these [quantities] are not determined by the radiation intensity at a certain frequency because in general many partial vibrations of the wave furnish a contribution to the radiation intensity at a certain frequency" (Planck 1958, Vol. I, p. 549). This is due to the fact that the measurable spectral intensity only yields "mean values" of the microscopic quantities of the field (amplitudes and phase constants). But from Eq. (3.17) it is clear that the resonator energy depends on these macroscopic quantities in the same fashion. If we now try to couple the resonator energy U with the corresponding spectral intensity J_v to evaluate the effect of the resonator on the field, we face an impasse: it is impossible to establish an unequivocal relationship between the Fourier coefficients of the resonator energy and those of the spectral intensity through the common microscopic quantities of the field. More specifically, Planck derives the following equations:

$$b_0 = \mathbf{B}_0^v + T \sum_n \eta_n \gamma$$

$$a_a = \mathbf{A}_a^v + \frac{\pi a}{\sigma n_0} \mathbf{B}_a^0 + T \sum_n \xi_{n,a} \gamma + T\pi a \sum_n \eta_{n,a} \gamma^2$$

$$b_a = \mathbf{B}_a^v - \frac{\pi a}{\sigma n_0} \mathbf{A}_a^0 + T \sum_n \eta_{n,a} \gamma - T\pi a \sum_n \xi_{n,a} \gamma^2.$$

The relationships between the mean quantities b_0, a_a, b_a and $\mathbf{B}_0^v$, $\mathbf{A}_a^v$, $\mathbf{B}_a^v$ call for the knowledge of the coefficients $\eta_{n,a}$, $\xi_{n,a}$, which in turn depend on the values of the amplitudes and phase constants of the individual components. Without these coefficients, we cannot evaluate the sums in the equations above and therefore we cannot calculate the coupling between resonator and spectral component: "a generally valid relation between the energy of the resonator and the intensity of the exciting wave does not exist, or [...] the energy of the resonator does not depend on the spectral intensity of the exciting wave only, rather [...] on the properties of the individual partial vibrations of this wave" (Planck 1958, Vol. I, p. 551). To get out of this deadlock it is necessary to make "the introduction of a limiting assumption

on the characteristics of the individual partial vibrations of the exciting wave." The
assumption put forward by Planck is the following:

$$\sum_n \eta_n \gamma_0 = \sum_n \xi_{n,a} \gamma_0 = \sum_n \eta_{n,a} \gamma_0 = \sum_n \xi_{n,a} \gamma_0^2 = \sum_n \eta_{n,a} \gamma_0^2 = 0 \qquad (3.19)$$

Equation (3.19) is the formal expression of the hypothesis of natural radiation
(HNR): there is a zero net difference between the microscopic amplitudes and phase
constants on the one hand and the Fourier coefficients on the other hand. It amounts
to saying that, when we come to calculate the effect of the resonator on the field, we
can forget about the behavior of individual amplitudes and phase constants. Their
disposition in the radiation is so disordered that their global behavior is completely
equivalent to the behavior of mean (macroscopic) quantities.

3.2.6 Molecular Chaos and Natural Radiation

Before going on with the analysis of Planck's theory, it is necessary to discuss the
HNR further. The reason is that this concept offers a nice opportunity to illuminate the
epistemic cooperation between the three dimensions of a theory. The HNR surely
has a representational content, because it concerns the way in which amplitudes
and phases behave in the electromagnetic field. However, the general picture of the
field that supports the HNR derives, as we have seen, from the Fourier series as a
technique to manipulate fields quantities. Furthermore, the HNR plays a crucial role
in the explanatory dimension of Planck's theory because it defines a relation between
macroscopic quantities and microscopic constituents of the field particularly suitable
for the kind of epistemic story Planck wants to tell about the cavity radiation. And, of
course, one should not forget that the selection and the organization of the concepts
and mathematical practices are influenced by that epistemic story.

To clarify the way in which the HNR serves the explanatory dimension, we have
to compare it with a notion in Boltzmann's kinetic theory: the hypothesis of molecular chaos. And, in turn, to understand the origin of molecular chaos, we have to
unravel the intricate debate that took place in the columns of *Nature* in 1894–95.
I have stated in Sect. 3.2.2 that in 1877, Boltzmann explained how the reversibility
objection could be reconciled with kinetic theory. For many years, that response
was considered acceptable, and kinetic theory developed on the tacit stipulation that
everybody agreed on that. This held until, in a letter to *Nature* on 25 October 1894,
E.P. Culverwell opened a Pandora's box (Culverwell 1894).[22] Commenting on the
second edition of H.W. Watson's book (Watson 1893), Culverwell cleverly condensed the reversibility objection by saying that if the molecular motion obeys the
laws of mechanics, then the number of processes going from an arbitrary state to the

---

[22]On the technical aspects of this debate see (Dias 1994). Culverwell had already objected in the
same vein in a previous paper (Culverwell 1890), which had apparently gone unnoticed.

equilibrium is equal to the number of those making the reverse path; therefore, it is impossible to prove a monotonic behavior of the function H by simply averaging the possible forms of Boltzmann's equation: this average must be zero. He concluded by candidly asking: "will some one say exactly what the H-theorem proves?"

Surprisingly, the specialists did not find themselves agreeing on a common answer, and the ensuing heated debate engaged physicists of the caliber of S.P. Burbury (1831–1911), G.H. Bryan (1864–1928), J. Larmor (1857–1942), Watson, and Boltzmann himself for months and closed without an explicit consensus. The reason is that, although it was apparent that kinetic theory required some kind of disorder assumption, at the core of this assumption lay a fundamental dualism that we can clarify using a distinction due to Jeans (1903). On the one hand, there was the *working form* of this assumption, namely the SZA. Everybody recognized that the SZA was absolutely necessary to derive Boltzmann's equation and the H-theorem. On the other hand, however, the opinions were seriously split on the *physical content* of this disorder assumption. The replies to Culverwell's provocative question can be sorted out into four categories.

Firstly, there were arguments that the time reversal is impossible because the H-theorem requires the SZA, and the SZA does not hold for the reversed collisions. This argument was brought about by Burbury, among others, who argued that the 'condition A', as he called the hypothesis of randomness in the collisions, cannot hold true in case of time reversal because the reversed collisions are no longer random (Burbury 1894).[23] This was a cheap way out of the reversibility objection: one stipulated a condition that was not satisfied by a time reversal. The obvious drawback was that the H-theorem would be reduced to a purely mathematical statement with unclear, if any, physical meaning. Incidentally, that was also the criticism that Boltzmann moved to Planck's first argument for irreversibility (see end of Sect. 3.2.3). But there were also more serious difficulties with this argument. For example, to explain irreversibility in a mechanical context by saying that 'natural' phenomena follow a unmechanical assumption seems to beg the question (Culverwell 1895a). More importantly, this argument might suggest the conclusion that if a state is disordered, then its time-reversal is ordered (Jeans 1903). But in this case, the argument is not a reply to Culverwell's objection that there are as many ordered state as disordered ones. As far as the state is concerned, the distinction between the original and its time-reversal is arbitrary: if A is the time-reversal of B, then the converse is also true. When Boltzmann happened to use this line of argument, he was very careful not to fall into this trap.

Secondly, there were arguments based on mechanical instability. It is true that the reversal is mechanically possible but, exactly like riding a bicycle backwards, it is a process that can very easily deviate from the initial trajectory to come back to equilibrium (Bryan 1894b, c, 1895; Burbury 1894). Occasionally, this idea went together

---

[23] A similar argument was proposed by (Bryan 1894a) in support of (Watson 1894).

with the remark that a system is never perfectly isolated and that external perturbations contribute to make the unstable states even more unstable. This argument was sporadically deployed by Boltzmann himself, although both the notion of instability and that of perturbation were only vaguely defined.

Thirdly, as in 1877, the mathematics of probability could be used to clarify the issue. Metaphorically, the behavior of the H-function resembles that of a tree: it is true that for each path that goes from the trunk to the tip of a branch the reverse path also exists, but if we are at some stage of the path, there are always more ways to go up than to go down. In this sense, the process is probabilistically one-directional: for each state S_0, the possible evolutions from S_0 toward equilibrium outnumber the evolutions toward non-equilibrium. Boltzmann deployed this argument in his replies to Culverwell, and apparently the latter was persuaded that this was the correct answer (Boltzmann 1895a, c, d; Culverwell 1895b). Successively, Boltzmann expanded on these thoughts in the notion of the H-curve (Boltzmann 1898a, 1909, Vol. III, pp. 629–637).

Finally, there is the concept of molecular chaos. Boltzmann elaborated upon this concept by collecting elements of this debate and adding to them a completely original trait. He anticipated this notion in the above-mentioned dispute on Kirchhoff's lectures,[24] but the more thorough discussion is in the first volume of the *Gastheorie* (Boltzmann 1898c, Sects. 3 and 6). To begin with, Boltzmann introduces the distinction between 'molar' and 'molecular'. The former refers to the presence of macroscopically perceptible dis-uniformities in the gas. For example, if the fast molecules are all disposed in a certain sub-volume, the system is molar-ordered.[25] But even if the system is perfectly uniform, it might still happen that some sort of order be at work at the microscopic level. Molecules could be arranged so as to collide in a specific, preconceived way: if that happens, the system is molecular-ordered, and the collision will follow a pattern decided in advance; if that does not happen, and the collisions follow the laws of probability (epitomized by the SZA), then the system is molecular-disordered. The gist of this distinction is that molecular chaos entails the use of probability laws: a state is ordered when it is purposefully prepared to evolve according to some of the theoretically possible paths:

> [I]f we choose the initial configuration on the basis of a previous calculation of the path of each molecule, so as to violate *intentionally* the laws of probability, then of course we can construct a persistent regularity or an almost molecular disordered distribution which will become molecular ordered at a particular time. (Boltzmann 1898c, Sect. 3), italic added

---

[24](Boltzmann 1895b), see Sect. 3.2.3.

[25] Thomas Kuhn has suggested that the molar/molecular distinction allowed Boltzmann to combine statistical and dynamical phraseology because it corresponds to the micro/macro distinction (Kuhn 1978, pp. 54–60). This terminology, however, was not new. J.J. Thomson, for instance, used a similar distinction to stress the uncontrollability of the molecular behavior (Thomson 1887). The molar/molecular distinction meant therefore something slightly different from the micro/macro divide, because the emphasis was preferably on the possibility of acting on microscopic states. It is precisely this point that Boltzmann wants to make with the concept of molecular chaos.

This aspect of 'conspiracy' inherent in the concept of order is the key to understanding Boltzmann's reply to the reversibility objection. For Boltzmann, the time reversal of one state always produces an ordered state because the reverse operation 'prepares' the state so as to contravene the laws of probability. Hence, to say that the SZA is not fulfilled by the reversed state means that the evolution of the system does not follow probability theory. As a consequence, the effect of reversal is as extended as far as the preparation:

> Consider any motion for which H decreases from time t_0 to time t_1. When one reverses all the velocities at time t_0, he would by no means arrive at a motion for which H must increase; on the contrary, H would probably still decrease. It is only when one reverses the velocities at time t_1 that he obtains a motion for which H must increase during the time interval $t_1 - t_0$ and even then H would probably decrease again after that, so that motions for which H continually remains very near to its minimum value are by far the most probable. (Boltzmann 1898c, Sect. 6)

There are reversed states that are still disordered in the sense that they produce an H-increasing evolution of the system, whereas H-decreasing evolution requires the system to be prepared in a particular way. To understand this point, remember the previous discussion about the distinction between ordered and disordered states. Let S_+ be an equilibrium state, and let S_- be its time-reversal. They are both equilibrium states, though neither of them is as such ordered or disordered. In fact, if no further information is known, there are no grounds on which to conclude whether S_+ will lead to equilibrium and S_- to non-equilibrium or the other way round. The laws of probability do not allow us to discriminate between these two cases. One can decide the question only if one has control of the process that brought about S_+ or S_-. This is Boltzmann's subtle point: once one has the control of such process, then one can produce an artificial, anti-thermodynamic evolution by inducing a molecular order in the system.[26]

But the more remarkable consequence is still to come. As claimed in the previous quotation, the H-theorem does not hold for systems that are prepared in an ordered state, but it is perfectly compatible with the probabilistic occurrence of any states. Let S_E be an equilibrium state and S_0 be a non-equilibrium state. Boltzmann's argument is that if S_E is followed by S_0 because of a special preparation, then the system is molecular-ordered. But the occurrence of a state is a matter of probability, and because S_0 has a nonzero probability to occur after S_E, molecular chaos is compatible with the process $S_E \rightarrow S_0$. In other words, *Boltzmann's molecular chaos does not prohibit anti-thermodynamic processes; it prohibits the time reversal as a ordering operation.*

This long and complex discussion is necessary to elucidate the historical and conceptual meaning of the HNR. Famously, Thomas Kuhn argued that, owing to Boltzmann's criticisms, Planck came to accept the kinetic approach on irreversibility

---

[26]Ultimately, Boltzmann's point amounts to distinguishing between equilibrium as a state described by Maxwell's distribution and as the end point of a mechanical trajectory. In other words, it concerns the epistemic surplus of two sets of symbolic practices: probabilistic tools and mechanical methods.

and tried to construe an electromagnetic analog of the molecular chaos, at the same time downplaying its probabilistic implications. The result is the HNR (Kuhn 1978, pp. 76–78).[27] This claim has two major implications. First, there is a sharp break after the third paper of the Pentalogy, and the aftermath of Boltzmann's criticism is that Planck finally accepted the necessity of introducing some sort of chaos assumption. The two parts of the Pentalogy should be separated by something like a gestalt switch. Second, from the fourth paper onwards, Planck looked to molecular chaos as the paradigmatic example of disorder assumption that had to be applied to his theory. This combination of discontinuity and paradigm is very Kuhnian, but historically questionable.

The main supporting fact for Kuhn's reconstruction is the similarity between the HNR and molecular chaos, a similarity stressed by Planck himself. My thesis is that Kuhn's reconstruction projects back to history a formal analogy, *and this is precisely what Planck wanted*. Planck's pronunciations about the relation between the HNR and molecular chaos come after 1900, i.e., after he accepted more elements of Boltzmann's approach, particularly combinatorics. At that point, he was more than willing to stress an underlying continuity in his theory, and the analogy between the HNR and molecular chaos came in handy. I will come back to this point with more textual evidence in Sect. 4.6.

Another point to bear in mind is that assumptions on the arrangement of the radiation components were already embodied in Planck's use of certain mathematical practices. As we have seen, he had made use of similar assumptions in the third paper.[28] Furthermore, the concept of molecular chaos was anything but an established notion. Boltzmann introduced it in the *Gastheorie* and never used it again. Jeans, in the paper mentioned above, complained that the molecular chaos was virtually useless to kinetic theory (Jeans 1903). Molecular chaos, in other words, was not the first place to look for an analogical adaption of Boltzmann's approach.

But the strongest point against Kuhn's reconstruction is that, beyond a superficial formal resemblance, the HNR differs radically from the concept of molecular chaos. When Boltzmann construes molecular chaos, he aims at grounding the probabilistic nature of the second law. He is ready to admit, as he does in the controversy with Zermelo, that highly improbable events, such as a return to non-equilibrium state, occur, but they occur because even highly improbable events have a nonzero probability. Therefore, molecular chaos allows for both probable and improbable events, but, of course, it makes also the former probable and the latter improbable. It is on this basis that Boltzmann can conclude that equilibrium is a matter of probability. This path is obviously barred for Planck: it collides head-on with the whole explanatory dimension of his theory. Accordingly, the HNR does a completely different thing:

---

[27]In effect, Kuhn regards the introduction of the HNR as the entering of statistics into radiation theory. Darrigol is more cautious and points out that Planck did not need to install Boltzmann's full package in his theory (Darrigol 1992, pp. 51–54).

[28]On this point see also (Seth 2010, pp. 119–126).

it cancels out all behaviors that do not conform to the second law of thermody-
namics. Assuming the HNR is tantamount to saying that anti-thermodynamic events
such as spontaneously jumping back from equilibrium to non-equilibrium simply do
not occur.[29] It is only in this way that the HNR can lead to a monotonic behavior
of entropy, as we will see in the next section. The HNR basically black-boxes the
microscopic state of the system and allows one to keep the analysis on the macroscopic state where irreversibility holds strictly. As a result, reversal and other odd
things do not happen: "if a resonator at a certain time is excited by a natural radiation
of variable intensity, the occurrence of the reverse process is excluded for all following times as long as the exciting wave conserves the properties of natural radiation"
(Planck 1958, Vol. I, p. 559). Planck would become even more explicit on this point
in the first edition of his *Vorlesungen über die Theorie der Wärmestrahlung* (Planck
1906, pp. 134–135). At that time, for strategic reasons, he was willing to minimize
the gulf between his approach and Boltzmann's and to stress a conceptual kinship
between radiation theory and kinetic theory. In a seldom-quoted paper published
in 1902, we find a telling account of the relation between gas theory and Planck's
particular view of radiation (Planck 1902, 1958, Vol. I, pp. 763–773). On commenting about new experiments about the nature of white light, Planck supported the
tradition of Gouy and Rayleigh according to which radiation is made up of many
closely entangled components. The use of this perspective requires, at some point,
an assumption on the disposition of these components, and here the analogy with
gas becomes manifest: "the amplitudes and the phases of the partial vibrations are
arranged in a completely irregular fashion, in particular two neighbor partial vibrations are in no relation whatsoever, just as what happens with the velocities of two
neighbor gas molecules" (Planck 1958, Vol. I, p. 766). After 1900, in his attempt
to tighten the link between radiation theory and kinetic theory, Planck sidestepped
one of the points on which he had erected his construction: contrary to molecular
velocities, amplitudes and phase have no independent meaning.

For these reasons, I suggest that we should view the introduction of the HNR
not as the product of an analogical import, but rather as the result of the epistemic
cooperation between the different dimensions of Planck's theory. It was motivated
by the multiple requirements deriving from the use of formal techniques and the
necessity of framing a certain epistemic story of equilibrium in cavity radiation.
Only by keeping in mind the mathematical practices Planck was using, his symbolic
codification of the radiation fields, and the demands of the explanatory dimension, we
can have a full picture of the role played by this hypothesis. The conceptual space of
the HNR was created with the concept of spectral decomposition of the intensity and
the analyzing resonator. Insisting on the necessity of a physically meaningful spectral
component, Planck engineered an interaction between intensity and resonator that
contained an intrinsic indetermination. Both the resonator and the spectral component
are made up of many microscopic constituents, and thus the problem requires some
additional hypotheses to be solved:

---

[29] I elaborate on the probabilistic consequences of this hypothesis in (Badino 2009); see also (Needell 1980).

Incidentally, this indetermination lies in the very nature of the things. For the physical problem has absolutely no determinate solution, as long as one knows nothing but the frequency and the damping constant of the resonator. [...] On the same token, this theory allows to establish nothing but the [frequency] and the [damping constant] of the resonator. Precisely in the gaps left by it, the hypothesis of natural radiation finds its place; were these gaps not there, then the hypothesis would be either superfluous or impossible because the process would be completely determined without it. (Planck 1958, Vol. I, p. 557)

The conceptual space, carefully tailored to permit the HNR, is justified by the inaccessibility and uncontrollability of the microscopic processes. This move paved the way for a concept of electromagnetic entropy and for the second argument for irreversibility.

3.2.7 Radiation Entropy

One immediate consequence of the HNR is that the effect of the resonator on the field can be straightforwardly calculated:

$$J'_\nu(t) - J_\nu(t) = -\frac{2}{\rho\nu}\frac{dJ_\nu(t)}{dt} + \frac{1}{\rho\sigma\nu^2}\frac{d^2J_\nu(t)}{dt^2}, \qquad (3.20)$$

where ρ is the damping constant of the analyzing resonator. Equation (3.20) expresses the difference of the secondary intensity $J'_\nu(t)$ with respect to the primary one. From this equation, Planck draws two important consequences. To begin with, a clearer definition of the stationary state is now possible. Planck shows that the intensity fluctuations of the secondary wave are always smaller that those of the primary wave. In the second place, Eq. (3.20) can be used to prove irreversibility. Planck remarks that "the most direct attribute of the irreversibility of a process lies in the proof of a function completely determined by the instantaneous state of the system, which has the properties of changing always in the same way, for instance decreasing, during the entire process" (Planck 1958, Vol. I, p. 554). Thus, caring very little about dimensional constraints, Planck introduces an 'entropy' for the resonator as a simple logarithmic function of the energy: $S_0 = \log U$. Analogously, he defines the radiation entropy as a slightly more complex function of the intensity. Calculating the composition of the two entropies and making use of Eq. (3.20), Planck arrives at a monotonically increasing expression.

This concept of electromagnetic entropy is very different from Wien's, as Planck is keen to stress. Wien's entropy is a thermodynamic function related to the system as a whole, while Planck's analogous concept is a property of the special electromagnetic process he has construed (Planck 1958, Vol. I, p. 536). This still ill-defined concept of entropy is the key ingredient of Planck's *second argument for irreversibility*. As the reader will recall, initially Planck tried to prove that one-directionality comes as a consequences of the boundary conditions of the process. After Boltzmann's criticism, Planck's new argument goes as follows. The process is irreversible because, assuming it fulfills the HNR, a function exists that only increases or possibly remains constant over time. But there is a missing link: the entropy function still needs a physical support. For the time being, its symbolic codification is exhausted by its being part of a set of mathematical practices. A new step was necessary.

3.2.8 The Systematization of Radiation Theory

It took Planck almost a full year to elaborate upon the physical ideas of the fourth paper. The result was the concluding installment of the Pentalogy presented on 1 June 1899 at the Berlin Academy (Planck 1899, 1958, Vol. I, pp. 560–600).[30] One distinctive feature of this last paper is the clear separation between the microscopic and the macroscopic parts of the theory. After quickly discussing the basic features of the resonator, Planck moves to the macroscopic treatment of radiation and arrives at a relation between the intensity and the resonator energy *via* HNR. Next, he generalizes the previous theory by taking into account the polarization state of the field. At the end, Planck finalizes his argument for irreversibility.

From a formal point of view, the most relevant improvement is the replacement of the Fourier discrete series with the Fourier integrals. This novelty was suggested by the mathematician Carl Runge[31] and enhanced the generality of Planck's approach, although it also involved some complications. The HNR is introduced in the transition from the microscopic to the macroscopic part, and it is elucidated by means of the distinction between rapidly and slowly varying quantities. Planck points out that microscopic quantities such as amplitudes, phase constants, and the dipole moment change very rapidly over time in a way that is out of our control and even perception. However, quantities such as the average energy of the resonator and the radiation intensity change relatively little in comparison with these quantities. The difference between the two realms of rapidly and slowly changing quantities is essential in Planck's theory. He is implicitly referring to the tradition of Clausius and Helmholtz, according to which macroscopic quantities are the empirical manifestation of many uncontrollable microscopic events. Although the microscopic events are disordered and complex, this chaotic variety is made of mutually compensating variations that, at the macroscopic level, provide a stable behavior.

Planck introduces the spectral decomposition in the same way as the previous paper[32] and introduces the HNR as a representation of the correct behavior of the

---

[30] Subsequently, Planck published a long paper in the *Annalen* that essentially reports the result of the fifth part of the Pentalogy with some small technical improvements (Planck 1900, 1958, Vol. I, pp. 614–667).

[31] Letter from Planck to Runge, 14 October 1898; see also (Planck 1958, Vol. I, p. 623).

[32] One formal difficulty that Planck does not succeed in overcoming is the dependence of the spectral component on the damping constant of the analyzing resonator. At the end, he is forced to drop it on the grounds that an acceptable spectral component cannot possibly depend on the features of the ideal apparatus used to measure it.

field: it is possible to replace the rapidly changing amplitudes and phase constants of the field with the slowly changing Fourier coefficients of the spectral intensity "without any perceptible error" (Planck 1958, Vol. I, p. 573).

The HNR entails a remarkable simplification in the calculation and leads immediately to a relation between the spectral intensity and the average energy of the resonator:

$$J_\nu = \frac{32\pi^2\nu^2}{3c^3}U_\nu. \tag{3.21}$$

In the last part of the paper, Planck deals with the thermodynamics of the radiation by using a formalism similar to Wien's: radiation is considered to be composed of rays traveling in straight lines along different directions (defined by a solid angle) and having different states of linear polarization. Therefore, the intensity must be decomposed into two principal intensities along two perpendicular polarization axes. Besides these innovations, the basic argument makes use of the familiar procedure of calculating energy conservation. The energy variation is the sum of the resonator energy and the integral of the density energy $u_\nu d\nu$ over all frequencies. All these quantities can be related to each other through the intensity by Kirchhoff's law, according to which the intensities associated with the waves absorbed and emitted by the resonator have the same ratio as the energy absorbed and emitted by the resonator itself. The energy density is related to the polarized intensity by the equation $u_\nu d\nu = \frac{8\pi K}{c}$. At the same time, combining an equation involving spectral intensity and polarized intensity with Kirchhoff's law, Planck links the average energy of the resonator with the polarized intensity $U = \frac{c^2}{\nu^2}K$. This result completes his analysis of the cavity radiation and lays the foundation for the concluding part of the paper, where irreversibility takes the stage.[33]

In the fifth paper, Planck finally systematizes the argument for irreversibility, which he had only outlined in the fourth communication. This time, Planck dwells upon the entropy function. In full analogy with energy, the total entropy of the system is the sum of the resonator entropy, summed over all resonators, and the entropy density of the radiation, integrated over all frequencies. The resonator entropy is defined as follows:

$$S = -\frac{U}{a\nu}\log\frac{U}{eb\nu}, \tag{3.23}$$

---

[33] It is interesting to note that Planck does not write explicitly the relation between energy density and average energy of the resonator:

$$u_\nu d\nu = \frac{8\pi\nu^2}{c^3}U_\nu, \tag{3.22}$$

which is usually associated with his electromagnetic radiation theory. To be sure, this relation can be easily derived from Planck's equations, but it is worth stressing that, at this stage, energy density and resonator energy are not the two macroscopic quantities he wants to relate. It is much more important for his irreversibility argument to work with the polarized intensity.

where a, b are constants, and e is the basis of the natural logarithm. Planck does not offer any justification for this definition: it is clear, as the argument goes, that he has obtained it by proceeding backwards from Wien's law. The definition of entropy density combines some formal similarities with the intensity and the resonator entropy. But the crucial point is the determination of the time-dependent process of entropy increase, which had been left implicit in the Eq. (3.20). From an energetic point of view, the process amounts to a transition $(\mathbf{K}_\nu, \mathbf{K}'_\nu) \rightarrow (\mathbf{K}''_\nu, \mathbf{K}'''_\nu)$ from a pair of intensities before the interaction with the resonator to the pair after the interaction. Understanding entropy as a quantity that can be ascribed to any physical process, Planck interprets the transformation from an entropic point of view as an analogous transition $(\mathbf{L}_\nu, \mathbf{L}'_\nu) \rightarrow (\mathbf{L}''_\nu, \mathbf{L}'''_\nu)$ where the $\mathbf{L}_\nu$'s are suitably defined 'entropy intensities' at different frequencies. This move enables him to view the variation of entropy as an entropic balance between two stages in the transformation. Planck applies to macroscopic entropy the technique used in kinetic theory for molecular energy. Exploiting the relation between entropy intensity and energy intensity already introduced by definition, Planck arrives at the following expression for the entropy variation:

$$
\frac{dS(t)}{dt} = \sum \frac{3c^2\sigma}{4\pi\nu} \int d\Omega \left(\mathbf{K} \log \frac{c^2\mathbf{K}}{e\nu^2 U_0} + \mathbf{K}' \log \frac{c^2\mathbf{K}'}{e\nu^2 U_0} - \mathbf{K}'' \log \frac{c^2\mathbf{K}''}{e\nu^2 U_0} \right.
$$
$$
\left. - \mathbf{K}''' \log \frac{c^2\mathbf{K}'''}{e\nu^2 U_0} \right).
$$

(3.24)

The similarity between this expression and the Eq. (3.5) used to prove the H-theorem is striking. The monotonic behavior follows immediately after some analytical manipulations. It is, in fact, a consequence of the mathematical relation between the quantities. In framing his new argument for irreversibility, Planck followed closely the argumentative pattern of kinetic theory. After all, as long as one accepts Boltzmann's equation, the H-theorem is nothing but an impeccable mathematical consequence. Planck was keen to maintain this latter trait and to configure entropy intensity in order to arrive at a similar formula. Somehow, the concept of radiation entropy is justified by this result. The resonator entropy, by contrast, is a different matter.

Wien had already defined the black-body radiation as the state of maximum entropy (see Sect. 2.3). Planck argues that this condition implies that "the variation of the total entropy [. . .] disappears for any infinitesimally small virtual variation of the state of the system" (Planck 1958, Vol. I, p. 590). The virtual variation Planck has in mind is the displacement of an infinitesimal quantity of energy from one resonator with frequency ν to another with frequency ν'. Then, the equilibrium condition and the energy conservation together entail that $\frac{1}{a\nu} \log \frac{U}{b\nu} = \frac{1}{a\nu'} \log \frac{U'}{b\nu'}$. Because the frequency is completely arbitrary, Planck concludes that the equilibrium between the resonators and the field depends on a single energy-dependent parameter:

$$\frac{1}{\Theta} = \frac{1}{av} \log \frac{U}{bv}. \tag{3.25}$$

The parameter Θ is the electromagnetic definition of temperature. Calculating U as a function of the temperature in the Eq. (3.25) and making use of the relations derived in the previous section, Planck immediately obtains Wien's radiation law in terms of frequency:

$$udv = \frac{8\pi bv^3}{c^3} e^{-\frac{av}{\Theta}} dv. \tag{3.26}$$

This equation is of course completely equivalent to the Eq. (2.6) as Planck demonstrates. In the very last section of the paper, Planck inquires about the thermodynamic consequences of his theory and the possibility of using some universal constants to construct a system of natural units of measures.[34] But the most interesting discussion for our purposes concerns the definition of the resonator entropy. Planck admits that there is no cogent argument in support of this definition, but he argues that any even slightly different expression would not yield Wien's law. There is no detailed proof of this statement, but the strategy is clear: to link the arbitrary entropy definition to the well-confirmed Wien's law as well as to the second principle of thermodynamics: "the [...] definition of radiation entropy and therewith the Wien energy distribution law are a necessary consequence of the application of the principle of the increase of entropy to the electromagnetic theory of radiation" (Planck 1958, Vol. I, p. 597). This link concludes Planck's radiation theory: now the entropy concept is anchored to an empirical result.

3.3 Overview of Planck's Radiation Theory

At the turn of the century, Planck could be reasonably satisfied with his radiation theory. Instead of adopting Wien's attitude of viewing heat radiation as the first step toward a detailed theory of matter and radiation, Planck appreciated its generality: heat radiation was the perfect site for a general theory of irreversible processes. His prime goal was not a derivation of Wien's law, but a particular form of explanation of the equilibrium in the radiation cavity. Planck's epistemic story relies completely on a strict interpretation of the second law of thermodynamics. His first argument for irreversibility was conceptually very simple: it is possible to set up an electromagnetic problem, with suitable equations and boundary conditions, such that the time-reversal of a solution is not a solution. To conceive of this argument, Planck relied mostly upon resources of electromagnetism, such as the asymmetry between advanced and retarded solutions of the wave equation and so on.

---

[34]For a discussion of this aspect see (Badino and Robotti 2001).

Boltzmann abruptly interrupted this dream with a crushing criticism. He showed beyond any doubt that Planck's alleged time-reversal was not at all genuine. Even worse, he showed that a real time reversal was always a solution of Planck's electromagnetic problem. At this point, Planck's program entered a phase of reorganization. On the established platform of electromagnetism and thermodynamics, some new elements had to be installed. He looked at the argumentative pattern used in kinetic theory. Previous theoreticians like Michelson and Wien had made a very opportunistic use of kinetic theory, but Planck's goal was different. The kinetic techniques and methods had an uncalled-for epistemic surplus: a statistical view of irreversibility that could not be reconciled with Planck's epistemic story. But, I have claimed, the epistemic surplus of mathematical practices is not always uniquely received. For Planck, statistical methods pointed to a view of the field as a bundle of rapidly changing microscopic quantities. This view, embodied in the HNR, paved the way for his thermodynamics of radiation as a technology for macroscopic quantities.

Planck integrated the kinetic argumentative pattern in his second argument for irreversibility. Adapting the formal structure of the H-theorem, Planck introduced an entropy function and showed that, by combining it with the equation derived from the HNR, a monotonic behavior of the system followed. Because the HNR cancelled out any recalcitrant evolutions, absolute irreversibility was safe. The only remaining stain in this second argument was the justification of the entropy function. Planck's solution was to link it to Wien's law. In fact, Planck did not give any derivation of the radiation law. This would require (1) a specification of the thermodynamically undetermined function of frequency and temperature and (2) a derivation of the exponential term. But Planck, from the very beginning, was not interested in demonstrating the radiation law: his stationary state was a uniform radiation that was already in thermal equilibrium. Planck's procedure to obtain Wien's law consists in the questionable virtual variation of the entropy.[35] After repeating that a resonator only interacts with a characteristic frequency, he assumed an unphysical transfer of energy between resonators with the sole goal of introducing the temperature. Once the temperature was introduced, through a procedure that is the exact contrary of Wien's, the radiation law follows analytically. There can be no doubt, however, that the form of the entropy function is justified by Wien's law and not the other way round. Thus, the most important effect of Planck's second argument is the close link between the entropy function, and ultimately his entire program, and Wien's law: they both stand or fall together.

The morphology of Planck's theory at the end of the Pentalogy is described in Fig. 3.2. The collapse of his first argument forces Planck to distinguish neatly between a microscopic part (on the left) and a macroscopic thermodynamics of radiation (on the right). Some of the symbolic practices travel from one side to the other, of course. In particular, the Fourier series play a crucial role as a formal representation of the complicated interaction of amplitudes and phases. While the first part of the Pentalogy presented a linear morphology from the electromagnetic premises to irreversibility,

---

[35]Criticisms against this procedure were raised in (Burbury 1902; Ehrenfest 1905); see below Sect. 4.7.1.

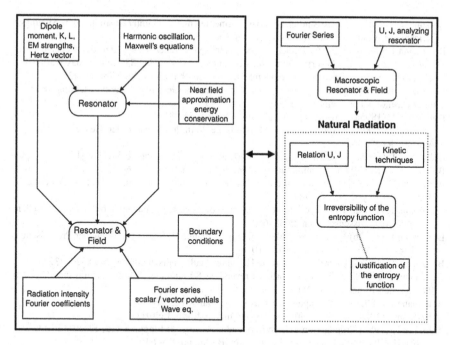

Fig. 3.2 Planck's theory at the end of the Pentalogy (1899)

here the structure in more complex. The leap to the thermodynamic part depends on the HNR, which works as an underlying hypothesis for the entire second argument. As we have seen, Planck does not linearly unfold a set of laws, but assembles his theory by combining the contribution of symbolic codifications, practices, and explanatory needs. This process of cooperation accounts for the numerous arbitrary steps, backwards calculations, and methodological posits one can find in the Pentalogy. The only dangling part is the justification of the entropy function. This is the part that, at the beginning of the new century, Planck sets out to fix.

References

Badino M (2009) The odd couple: Boltzmann, Planck and the application of statistics to physics (1900–1913). Annalen der Physik 18(2–3):81–101

Badino M (2011) Mechanistic slumber vs. statistical insomnia: the early phase of Boltzmann's H-theorem (1868–1877). Eur Phys J H 36:353–378

Badino M, Robotti N (2001) Max Planck and the constants of nature. Ann Sci 58:137–162

Bjerknes V (1891a) Über den zeitlichen Verlauf der Schwingungen im primären Hertz'schen Leiter. Annalen der Physik 44:513–526

Bjerknes V (1891b) Über die Erscheinung der multiplen Resonanz electriscer Wellen. Annalen der Physik 44:92–101

Boltzmann L (1872) Weitere Studien über das Wärmegleichgewicht unter Gasmolekülen. Sitzungs-berichte der Akademie der Wissenschaften zu Wien 66:275–370

Boltzmann L (1877) Über die Beziehung zwischen dem zweiten Hauptsatze der mechanischen Wärmetheorie und der Wahrscheinlichkeitsrechnung respective den Sätzen über das Wärmegle-ichgewicht. Sitzungsberichte der Akademie der Wissenschaften zu Wien 76:373–435

Boltzmann L (1894) Über den Beweis des Maxwellschen Geschwindigkeitsverteilungsgesetzes unter Gasmolekülen. Annalen der Physik 53:955–958

Boltzmann L (1895a) Erwiderung an Culverwell. Nature 51:581

Boltzmann L (1895b) Nochmals das Maxwellsche Verteilungsgesetz der Geschwindigkeiten. Annalen der Physik 55:223–224

Boltzmann L (1895c) On certain questions of the theory of gases. Nature 51:413–415

Boltzmann L (1895d) On the minimum theorem in the theory of gases. Nature 52:211

Boltzmann L (1896) Entgegnung auf die wärmetheoretischen Betrachtungen des Hrn. E. Zermelo. Annalen der Physik 57:773–784

Boltzmann L (1897a) Über irreversible Strahlungsvorgänge I. Sitzungsberichte der Preussischen Akademie der Wissenschaften 2:660–662

Boltzmann L (1897b) Über irreversible Strahlungsvorgänge II. Sitzungsberichte der Preussischen Akademie der Wissenschaften 2:1013–1018

Boltzmann L (1898a) Über die sogenannte H-Kurve. Mathematische Annalen 50:325–332

Boltzmann L (1898b) Über vermeintlich irreversible Strahlungsvorgänge. Sitzungsberichte der Preussischen Akademie der Wissenschaften 1:182–187

Boltzmann L (1898c) Vorlesungen über Gastheorie. Barth, Leipzig

Boltzmann L (1909) Wissenschaftliche Abhandlungen. Barth, Leipzig

Bremmer H (1958) Propagation of electromagnetic waves. In: Flügge S (ed) Electric Fields and Waves, Handbuch der Physik, vol 15. Springer, Berlin, pp 423–639

Brown HR, Myrvold W, Uffink J (2009) Boltzmann's H-theorem, its discontents, and the birth of statistical mechanics. Stud Hist Philos Mod Phys 40:174–191

Bryan GH (1894a) The kinetic theory of gases I. Nature 51:176

Bryan GH (1894b) The kinetic theory of gases II. Nature 51(1311):152

Bryan GH (1894c) The kinetic theory of gases III. Nature 51(1312):176

Bryan GH (1895) The assumption in Boltzmann's minimum theorem. Nature 52:29–30

Buchwald J (1998) Reflections on Hertz and the Hertzian dipole. In: Baird D, Hughes RIG, Nordmann A (eds) Heinrich Hertz: classical physicists. Modern Philosopher, Kluwer Academic, London, pp 269–280

Burbury SH (1894) Boltzmann's minimum function. Nature 51(1308):78–79

Burbury SH (1902) On irreversible processes and Planck's theory in relation thereto. Philos Mag 3(14):225–240

Culverwell EP (1890) Note on Boltzmann's Kinetic theory of gases, and on Sir W. Thomson's Address to Section A, British Association, 1884. Philos Mag 30(182):95–99

Culverwell EP (1894) Dr. Watson's proof of Boltzmann's theorem on permanence of distributions. Nature 50(1304):617

Culverwell EP (1895a) Boltzmann's minimum theorem. Nature 51(1315):246

Culverwell EP (1895b) Professor Boltzmann's letter on the kinetic theory of gases. Nature 51(1329):581

Darrigol O (1992) From c-numbers to q-numbers. The classical analogy in the history of quantum theory. University of California Press, Berkeley

Dias PMC (1994) "Will someone say exactly what the H-theorem proves?" A study of Burbury's condition A and Maxwell's proposition II. Arch Hist Exact Sci 46(4):341–366

Ehrenfest P (1905) Über die physikalischen Voraussetzungen der Planck'schen Theorie der irre-versiblen Strahlungsvorgänge. Sitzungsberichte der Akademie der Wissenschaften zu Wien 114:1301–1314

Ehrenfest P, Ehrenfest T (1911) The conceptual foundations of the statistical approach in mechanics. Dover, New York

Essex EA (1977) Hertz vector potentials of electromagnetic theory. Am J Phys 45(11):1099–1101

Gouy LG (1886) Sur le mouvement lumineux. Journal de Physique Théorique et Appliquée 5:354–362

Hertz H (1889) Die Kräfte electrischer Schwingungen, behandelt nach der Maxwell'schen Theorie. Annalen der Physik 36(1):1–22

Höflechner W (1994) Ludwig Boltzmann. Akademische Druck und Verlaganstalt. Leben und Briefe, Graz

Jeans JH (1903) The kinetic theory of gases developed from a new standpoint. Philos Mag 5(30):597–620

Klein MJ (1973) The development of Boltzmann's statistical ideas. Acta Phys Austriaca Suppl 10:53–106

Kuhn T (1978) Black-body theory and the quantum discontinuity, 1894–1912. Oxford University Press, Oxford

Loschmidt J (1876) Über den Zustand des Wärmegleichgewichtes eines Systems von Körpern mit Rücksicht auf die Schwerkraft. Sitzungsberichte der Akademie der Wissenschaften zu Wien 73:128–142

Maxwell J (1867) On the dynamical theory of gases. Philos Trans Roy Soc Lond 157:49–88

Maxwell J (1871) Theory of heat. Longmans, Green and Co., London

Needell A (1980) Irreversibility and the failure of classical dynamics: Max Planck's work on the quantum theory, 1900–1915. PhD thesis, University of Michigan, Ann Arbor

Planck M (1879) Über den zweiten Hauptsatz der mechanischen Wärmetheorie. Ackermann, München

Planck M (1891) Allgemeines zur neueren Entwicklung der Wärmetheorie. Zeitschrift für physikalische Chemie 8:647–656

Planck M (1894) Antrittsrede, gehalten am 28. Juni 1894 zur Aufnahme in die Akademie. Sitzungsberichte der Preussischen Akademie der Wissenschaften 2:641–644

Planck M (1895) Über den Beweis des Maxwellschen Geschwindigkeitsverteilungsgesetzes unter den Gasmolekülen. Annalen der Physik 55:220–222

Planck M (1896) Absorption und Emission electrischer Wellen durch Resonanz. Annalen der Physik 57(1):1–14

Planck M (1897a) Über electrische Schwingungen, welche durch Resonanz erregt und durch Strahlung gedämpft werden. Annalen der Physik 60:577–599

Planck M (1897b) Über irreversible Strahlungsvorgänge. 1. Mitteilung. Sitzungsberichte der Preussischen Akademie der Wissenschaften 1:57–68

Planck M (1897c) Über irreversible Strahlungsvorgänge. 2. Mitteilung. Sitzungsberichte der Preussischen Akademie der Wissenschaften 2:715–717

Planck M (1898a) Über irreversible Strahlungsvorgänge. 3. Mitteilung. Sitzungsberichte der Preussischen Akademie der Wissenschaften 1:1122–1145

Planck M (1898b) Über irreversible Strahlungsvorgänge. 4. Mitteilung. Sitzungsberichte der Preussischen Akademie der Wissenschaften 2:449–476

Planck M (1899) Über irreversible Strahlungsvorgänge. 5. Mitteilung. Sitzungsberichte der Preussischen Akademie der Wissenschaften 1:440–480

Planck M (1900) Über irreversible Strahlungsvorgänge. Annalen der Physik 1:69–122

Planck M (1902) Über die Natur des weissen Lichtes. Annalen der Physik 7:390–400

Planck M (1906) Vorlesungen über die Theorie der Wärmestrahlung. Barth, Leipzig

Planck M (1958) Physikalische Abhandlungen und Vorträge. Vieweg, Sohn, Braunschweig

Rayleigh JWS (1889) On the character of the complete radiation at a given temperature. Philos Mag 27:460–469

Rowland HA (1884) On the propagation of an arbitrary electro-magnetic disturbance, on spherical waves of light and the dynamical theory of diffraction. Am J Math 6(4):359–381

Schuster A (1894) On interference phenomena. Philos Mag 37:509–545

Seth S (2010) Crafting the quantum. Arnold sommerfeld and the practice of theory, 1890–1926. MIT Press, Cambridge

Thomson JJ (1887) Some applications of dynamical principles to physical phenomena. Part II. Philos Trans Roy Soc Lond 178:471–526

Watson HW (1893) A treatise on the Kinetic theory of gases, 2nd edn. Clarendon Press, Oxford

Watson HW (1894) Boltzmann's minimum theorem. Nature 51(1309):105

Wien W (1893) Die obere Grenze der Wellenlängen, welche in der Wärmestrahlung fester Körper vorkommen können; Folgerungen aus dem zweitem Hauptsatz der Wärmetheorie. Annalen der Physik 49:633–641

Wien W (1909) Theorie der strahlung. In: Sommerfeld A (ed) Encyklopädie der mathematischen Wissenschaften, vol 3. Teubner, Leipzig, pp 282–357

Zermelo E (1896) Über einen Satz der Dynamik und die mechanische Wärmetheorie. Annalen der Physik 57:485–494

Chapter 4
Deconstructing Planck

Abstract This chapter concludes the analysis of Planck's theory of radiation, and discusses the fateful leap to the quantum hypothesis. In the Pentalogy, Planck had bounded his theory tight with Wien's radiation law. Ironically, he wasn't looking for any proof of the law to begin with. But it was necessary to buttress his definition of entropy in order to conclude his second argument for irreversibility. In March 1900, he tightened this link even more. However, some month later, Rubens' experiments showed that Wien's law was untenable. By carefully analyzing Planck's combinatorial procedure between October and December 1900, I argue that he took a noncommittal stance respect to the question whether the quantization concerned the nature of the resonators or a way to distribute a continuum of energy. The reason was that Planck's symbolic practices are compatible with two radically different combinatorial models both to be found in Boltzmann's 1877 work. This plurality of representation is another case of the same underdetermination that had been treated with the HNR. Thus, although Planck did introduce the quantum hypothesis in 1900, this hypothesis remained in epistemic isolation until a more thorough exploration of Planck's theory allowed the physics community to appreciate its nonclassical meaning.

Keywords Thermodynamics · Planck's radiation law · Combinatorics · Probability · Quantum hypothesis · Electromagnetism · Statistics

4.1 Development and Crisis

4.1.1 Nothing Quiet on the Experimental Front

The natural question is: how strong was the empirical support for Wien's law? In effect, experiments started to cast shadows over the law well before Planck's Pentalogy was completed. Otto Lummer (1860–1925) and Ernst Pringsheim (1859–1917), who were competing with Paschen on the experimental investigations on the blackbody, communicated to the German Physical Society minor deviations from Wien's

© The Author(s) 2015
M. Badino, *The Bumpy Road*, SpringerBriefs in History
of Science and Technology, DOI 10.1007/978-3-319-20031-6_4

law at long wavelengths as early as February 1899 (Lummer and Pringsheim 1899a).[1] At the end of the summer, Rubens reached a similar conclusion working on the residual rays in the region of 1030 K and wavelength around 22μ, but he remained convinced that the slight deviations were due to the apparatus and had "nothing to do" with Wien's law (Rubens 1899, p. 588). On 19 September, Planck presented his theory at the Meeting of the German Scientists at Munich. On that occasion, he received some verbal remarks from Boltzmann immediately implemented in the final paper for the *Annalen*. Before he could send off the paper, though, Lummer and Pringsheim produced a second battery of experiments (Lummer and Pringsheim 1899b) in which the deviations were still tiny, but apparently systematic, at temperatures as high as 1646 K (Fig. 4.1).

Planck took due notice of these experimental difficulties in a footnote at the end of his paper, but he was not worried (Planck 1958, Vol. I, p. 662). Not all theoreticians agreed with this opinion. At an eventful meeting of the German Physical Society on 2 February 1900, Max Thiesen took very seriously the experimental outcomes and put forward a family of possible distribution functions that encompassed all the available data (Thiesen 1900). However, Thiesen was unable to provide a theoretical derivation for the family of functions. At the same meeting, Lummer and Pringsheim reported on their most recent experiments (Lummer and Pringsheim 1900).[2] After investigating the behavior of a black-body between 12 and 18μ and in a temperature interval between 85 and 1800 K, they concluded that "the Wien-Planck spectral equation does not represent the black radiation measured by us" (Lummer and Pringsheim 1900, p. 171). Deviations were systematic both at increasing wavelength at constant temperature and at increasing temperature at constant wavelength (Fig. 4.2).

Thus, at the beginning of 1900, experimenters had pronounced their verdict against Wien's law. But for Planck, there was still no need to worry too much. Although it was fairly clear that Wien's law was imperfect, there was also the strong feeling that this imperfection called only for a minor adjustment. Both Thiesen and Lummer and Pringsheim pointed out that the experimental curves could have been better represented by slightly changing the exponential function. Planck was much more concerned about the theoretical flaws of the Pentalogy. His precarious attempt to justify the entropy definition as the simplest one yielding Wien's law was immediately criticized by Lummer and Pringsheim in their November paper (Lummer and Pringsheim 1899b, p. 225). They asked for a compelling proof that Wien's law was compatible with only one possible entropy formula, a proof that Planck could not provide. Instead, he followed a different, and more familiar, route.

---

[1] For a thorough discussion of the experimental works see (Kangro 1970).

[2] The text of the communication was published very late, even after Lummer and Pringsheim's talk at the September Meeting of the German Scientist at Aachen. On the experimental research on the black-body radiation about 1900 see (Hoffmann 2001).

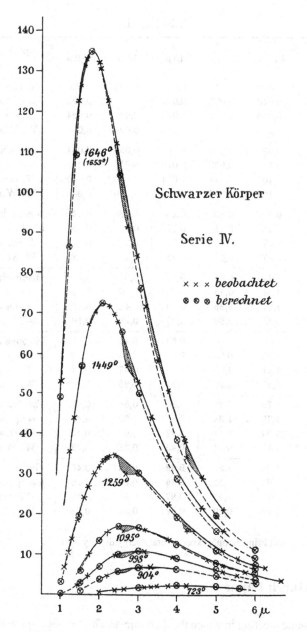

Fig. 4.1 Lummer and Pringsheim's experimental *curve* of the energy distribution for a black-body

Tabelle III.

Abs. Temp.	12,3 μ	13,25 μ	15 μ	16,5	17,9 μ	Spectral-gleichung
287°	0,040	0,038	0,030	0,023	0,020	beobachtet
	0,040	0,037	0,031	0,026	0,022	$\mu = 4$; $\nu = 1,3$
	0,040	0,087	0,032	0,028	0,024	THIESEN
	0,039	0,086	0,030	0,026	0,022	W. WIEN
373°	0,099	0,088	0,066	0,050	0,041	beobachtet
	0,104	0,090	0,070	0,055	0,045	$\mu = 4$; $\nu = 1,3$
	0,107	0,095	0,074	0,060	0,049	THIESEN
	0,107	0,094	0,074	0,060	0,049	W. WIEN
700°	0,49	0,40	0,26	0,18	0,14	beobachtet
	0,51	0,41	0,27	0,18	0,15	$\mu = 4$; $\nu = 1,3$
	0,57	0,45	0,31	0,22	0,17	THIESEN
	0,47	0,36	0,23	0,16	0,12	W. WIEN
1095°	1,11	0,87	0,52	0,37	0,27	beobachtet
	1,11	0,86	0,56	0,40	0,29	$\mu = 4$; $\nu = 1,3$
	1,24	0,95	0,60	0,42	0,30	THIESEN
	0,85	0,63	0,38	0,26	0,18	W. WIEN
1200°	1,29	1,00	0,59	0,42	0,31	beobachtet
	1,28	0,99	0,63	0,45	0,33	$\mu = 4$; $\nu = 1,3$
	1,41	1,08	0,68	0,47	0,34	THIESEN
	0,93	0,69	0,42	0,28	0,19	W. WIEN
1492°	1,78	1,35	0,80	0,56	0,41	beobachtet
	1,75	1,34	0,85	0,60	0,43	$\mu = 4$; $\nu = 1,3$
	1,87	1,41	0,87	0,60	0,43	THIESEN
	1,11	0,81	0,48	0,31	0,22	W. WIEN
1650°	1,96	1,52	0,92	0,63	0,46	beobachtet
	2,01	1,53	0,96	0,67	0,48	$\mu = 4$; $\nu = 1,3$
	2,11	1,58	0,97	0,66	0,48	THIESEN
	1,23	0,89	0,52	0,34	0,23	W. WIEN

Fig. 4.2 Lummer and Pringsheim's experimental energy density

4.2 The March Paper

At the very same meeting in which the theoretician Thiesen together with the experimenters Lummer and Prinsheim undermined the trust in Wien's law, Planck discussed an alternative way to support his entropy formula (Planck 1900a, 1958, Vol. I, pp. 668–686). His paper was eventually published in the *Annalen* the following March; therefore, I refer to it as the 'March paper'.[3] The main idea was to treat the radiation

---

[3] Kuhn suggests that the argument of the paper might have been somehow prepared by Thiesen's general analysis (Kuhn 1978, pp. 94–95).

field as a thermodynamic system and to add special assumptions to derive an expression for the entropy. However, before discussing Planck's argument, it is interesting to dwell briefly upon the introduction of the paper. In the first place, Planck downplays the alleged empirical confutations of Wien's law. Opposing Friedrich Paschen's measurements (Paschen 1899) to Lummer's and Pringsheim's, Planck concludes that there is room for a sensible improvement, but no need for a dramatic change: "the pending questions among the observers represent to me an incentive to order and criticize sharply the theoretical presuppositions that lead to the [...] entropy expression and that must be somehow changed in case the Wien distribution law will be found not valid in general" (Planck 1958, Vol. I, p. 669).

One second important point concerns a further comparison between radiation theory and kinetic theory. Albeit long, I quote Planck's comment entirely, because it will become very important later on:

> A further objection against this theory can stem from the fact that the irreversibility of the radiation processes and the concept of radiation entropy is derived from the analysis of one and only one resonator, whereas in gas theory it is use and wont to assume that the emergence of irreversibility and the definition of entropy are possible only with a large number of molecules. This objection can be further enhanced by the consideration that the principle of disorder, on which any kind of irreversibility seems to rely, enters gas theory in a very different circumstance. In gases several ponderable molecules originate disorder through the irregularity of their position and velocity; on the contrary, in a cavity full of radiation it is several bundles of rays that form the entropy through their irregularly changing frequencies and amplitudes. In the oscillations of a single resonator this irregularity is as much expressed as in the free space radiation. Since, while in gas theory the *vis viva* of a single molecule is only a negligible fraction of the kinetic energy of the smallest portion of gas and has no independent meaning when isolated, in radiation theory the energy of a single resonator has the same order of magnitude of the energy of free radiation contained in a space much larger than the dimensions of the resonator itself. Accordingly, the stationary oscillation of a resonator in a stationary field does not represent an elementary process, that is a simple sinus oscillation with constant amplitude and phase [...], but rather it consists of a superposition of very many individual oscillations with almost equal periods and constant amplitudes and phases. [...] In this case, one can speak of a disorder as well as of an entropy and a temperature of the resonator. (Planck 1958, Vol. I, pp. 673–674)

Here, Planck is, for the first time, explicit in relating his procedure with usual kinetic theory. He states that the apparently intuitive analogy between resonator and gas molecule—an analogy based on their alleged materiality—is misleading: the motion of a molecule is an elementary process, while the oscillation of the resonator is a collection of many elementary amplitudes and phases. This is the reason why Planck's radiation theory works very well with a single resonator only. This conceptual link between the two theories will be changed by the introduction of combinatorics (see Sect. 4.5).

Next, Planck develops his argument to justify the entropy function. He begins with a system in stationary state and investigates what happens when the energy is slightly modified. Let us assume that the energy of the stationary resonator has been changed to $U = U_0 + \Delta U$, where U_0 is the value in equilibrium and ΔU a very small (positive or negative) quantity. The system will tend to restore the equilibrium state, i.e., the entropy will move back to the maximum value. Now, the entropy variation is simply the difference between the entropy acquired by the resonator and lost by the radiation. Here, Planck makes a surprising move: he expands in Taylor series the entropy variation for the non-equilibrium state. This expansion yields the second derivative of the entropy function d^2S/dU^2, a crucial quantity for evaluating the entropy variation. It turns out that such variation depends on the entropy as a second order effect $dS_t = dU \cdot \Delta U \cdot \frac{3}{5} \frac{d^2S}{dU^2}$. The second derivative of the entropy is a quantity that never appears in thermodynamics. In this particular case, it is a consequence of the concept of electromagnetic entropy that combines thermodynamic codification with the typical analytical symbolic practices of a field theory.

We can speculate that Planck was led to the introduction of the second derivative by another inter-dimensional cooperation. The entropy must yield the derivation of the famous exponential term in Wien's law. Now, the first derivative is used to introduce the temperature through the thermodynamic relation $\frac{dS}{dU} = \frac{1}{T}$. One moment's reflection reveals that the exponential factor can be found by integrating an expression such as $\frac{c}{U}$ with c constant, therefore the second derivative had to be related to temperature *via* such a function.

The total entropy is always positive if:

$$dS_t = -dU \cdot \Delta U \cdot f(U),$$

where $f(U)$ is a positive function of the energy. This result follows from the principle of increase of entropy only. To understand more about the function $f(U)$, Planck assumes that there are n independent resonators in the cavity. Because both energy and entropy are extensive quantities, the variation of the total entropy is $d(nS_t)$, and this entails:

$$dU_n \cdot \Delta U_n f(U_n) = n[dU \cdot \Delta U \cdot f(U)], \tag{4.1}$$

where $U_n = nU$, $\Delta U_n = n\Delta U$, and $U_n = nU$. Manipulating this equation makes it possible to show that $nf(nU) = f(U)$, and this yields the condition on the second derivative of the entropy $\frac{d^2S}{dU^2} = -\frac{\alpha}{U}$, where α is a positive constant. This was exactly the relation needed. By twice integrating this expression one can immediately arrive at an entropy formula completely equivalent to Eq. (3.23).[4]

The ingredients of this argument could not be any simpler: the principle of increase of entropy, some basic analytical techniques, and the property of extensivity. But the simplicity and the generality concealed a threat. Now the morphology of Planck's theory has changed again, and the bound between Planck's theory and Wien's law was tighter than ever (Fig. 4.3). Planck had committed himself to an extremely general thermodynamic argument that led to a very specific form for the second derivative of the entropy and to Wien's law. All of a sudden, Wien's law was the cornerstone of the entire architecture, because without that form of the second derivative there was

---

[4]The dependence on the frequency comes from the application of the displacement law, which, in this case, states that the entropy must be a function of the ratio U/ν.

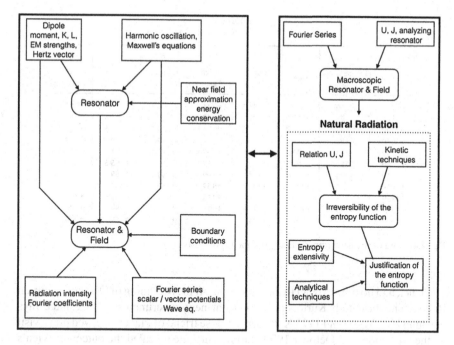

Fig. 4.3 Planck's theory in the March paper (1900)

no entropy, and therefore, no argument for irreversibility. The conceptual structure of the theory was now so rigid that the whole program depended essentially on a single second derivative.

4.3 Crisis

Compelling as it seemed, Planck's argument did not go unquestioned. At the International Physical Conference held in Paris in August 1900, Wilhelm Wien delivered a thorough analysis of the state of the radiation theory. On closing his paper, Wien criticized Planck's March argument. He posited that, according to this argument, the entropy formula was a necessary consequence of the second principle applied to *more than one* resonator. But, as Planck had insisted in the introduction of his paper, the resonators are independent, hence irreversibility should follow from the action of one resonator only. This logical difficulty seemed to undermine the whole argument: either the conclusion or one of the premises must be false (Wien 1900, p. 40).[5]

---

[5]At the same meeting, Lummer also objected to Planck's theory that there was no proof that many resonators would behave like one with regards to the entropy increase (Lummer 1900, p. 92).

Tabelle II.

Reststrahlen von Steinsalz, $\lambda = 51.2\,\mu$.

Temperatur in Celsius-Graden t	Absolute Temperature T	E beob.	E nach WIEN	E nach THIESEN	E nach RAYLEIGH	E nach LUMMER u. JAHNKE	E nach PLANCK
− 273	0	—	−121.5	− 44	− 20	− 27	− 23.8
− 188	85	− 20.6	−107.5	− 40	− 19	− 24.5	− 21.9
− 80	193	− 11.8	− 48.0	− 21.5	− 11.5	− 13.5	− 12.0
+ 20	293	0	0	0	0	0	0
+ 250	523	+ 31.0	+ 63.5	+ 40.5	+ 28.5	+ 31	+ 30.4
+ 500	773	+ 64.5	+ 96	+ 77	+ 62.5	+ 65.5	+ 63.8
+ 750	1023	+ 98.1	+118	+106	+ 97	+ 99	+ 97.2
+1000	1273	+132	+132	+132	+132	+132	+132
+1250	1523	+164.5	+141	+154	+167	+165.5	+166
+1500	1773	+196.8	+147.5	+175	+202	+198	+200
+ ∞	∞	—	+194	+ ∞	+ ∞	+ ∞	+ ∞

Fig. 4.4 Data of the isochromatic energy of the residual rays

Planck had no time to react to this objection. At the beginning of October, Heinrich Rubens and Friedrich Kurlbaum carried out new experiments with residual rays reaching $\lambda = 51.2\mu$ and 1773 K. Rubens, a close friend of Planck's, payed him a visit on the afternoon of 7 October 1900, during which he revealed the outcome: Wien's law broke down completely as the product λT reached a certain value. Because this also happened with formulae similar to Wien's, the correct distribution law had to be dramatically different (Hettner 1922). Working on Rubens' experimental data, Planck calculated by interpolation the correct law and communicated it to the Berlin Academy on 19 October (Planck 1900b, 1958, Vol. I, pp. 687–689). Rubens, who had waited for Planck to check his calculation, confirmed officially the formula five days later (Rubens and Kurlbaum 1900); Fig. 4.4.

That was the end of Wien's law. Planck could not neglect Rubens' and Kurlbaum's experiments as he had done with Lummer's and Pringsheim's for a crucial reason: working backward from the correct distribution law, he had found out that the second derivative of the entropy was hopelessly at odds with that of the March paper. The thermodynamic argument had made Planck's program so dependent on a specific form of the second derivative that a different form would have troubled it badly. And this was precisely what happened.

4.4 A New Radiation Law

The decisive phase in the emergence of the quantum hypothesis covers three months and three very different papers. In particular, the first two communications, (Planck 1900b) and (Planck 1900c, 1958, Vol.I, pp. 698–706) were meant to be read before the German Physical Society and properly supplemented by discussion, while the third paper was a self-contained publication for *Annalen der Physik* (Planck 1901, 1958,

Vol.I, pp. 717–727). Events followed one another at a frantic pace, so it is pointless to suppose a sort of linear route to the final result. The published material at our disposal—which is in turn very intricate indeed—is the superficial manifestation of a bumpy road. An arrangement of this material along a straightforward argumentative line would misrepresent the complex process that led Planck to the final breakthrough. Planck worked in parallel all the time, trying to combine the different dimensions of his theory. Traces of this 'multitasked' work can be found in the papers. Hence, I think that a sounder approach to this material is a deconstructive one: to isolate the different strands of Planck's reasoning and their relative epistemic contribution to the final outcome. The resulting narrative does not unravel the development of an argument, but rather the succession of stages in which Planck progressively took crucial decisions on different parts of the theory.

To begin with, Planck had to face the experimental debunking of Wien's law. After Rubens' personal communication, Planck realized that the substantial part of his March paper could not be maintained. Apparently, he first attempted to modify that argument to give a thermodynamic foundation to the correct distribution law. The October paper bears traces of this attempt, which failed very soon. Between October and December, Planck looked for another path. The second stage consisted in the decision to adopt combinatorial arguments. In this phase, Planck worked on the formal properties of the entropy definition in parallel with Boltzmann's 1877 theory and used portions of his March argument. Most probably he arrived at a plausible combinatorial formula before having a clear idea of its meaning. This brought him to the third stage, the construction of a combinatorial model. Again, he worked in parallel with Boltzmann's theory in order to find out the most effective adaptation of Boltzmann's combinatorial procedure to a system of oscillators. In this phase, Planck's reasoning concerns chiefly the formal properties of the combinatorial formula: he reads its epistemic surplus in the same way as he had done with other probabilistic methods. For Planck, combinatorics was a set of techniques to manipulate partial information and called for a black-boxing of the microscopic state. This reading was buttressed by the discovery that the combinatorial formalism was compatible with two different accounts on the nature of radiation. This state of affairs encouraged Planck to take a noncommittal stance regarding the entire combinatorial procedure, including the quantum hypothesis.

Let us follow now this argument in detail. Working on the experimental data provided by Rubens, and later confirmed by Paschen, Planck found out that the radiation intensity as a function of the temperature and the wavelength was represented by the new law:

$$K(\lambda, T) = \frac{c_1}{\lambda^5} \frac{1}{e^{-\frac{c_2}{\lambda T}} - 1}, \tag{4.2}$$

where c_1, c_2 are universal constants. Apparently, the change in the formula is small, but its consequences are remarkable. First, because if λT is large (high temperature, long wavelengths), the exponential term can be expanded in Taylor series as $e^{-\frac{c_2}{\lambda T}} = 1 + \frac{c_2}{\lambda T}$, and a substitution gives a formula that, contrary to Wien's, increases steadily with temperature.[6] Second, to take into account the very different behaviors at the two extremes of the spectrum, the second derivative must be substantially modified: $\frac{d^2 S}{dU^2} = \frac{\alpha}{U(\beta+U)}$. As a preliminary attempt to shed light upon the problem, Planck tries to maintain the same morphology of Fig. 4.3 and traces back the new radiation law to his March argument. If the second derivative has to be changed, then some part of the argumentative chain leading to it has to be changed as well. Planck focuses on the Eq. (4.1), about which he comments in a rather enigmatic fashion[7]:

> I could consider the possibility, even if it would not be easily understandable and in any case would still be difficult to prove, that the expression on the left-hand side [of the Eq. (4.1)] would not have the general meaning which I attributed to it earlier, in other words: that the values of U_n, dU_n and ΔU_n are not by themselves sufficient to determine the change of entropy under consideration, but that U itself must also be known for this. (Planck 1958, Vol. I, p. 688)

Planck was implicitly referring to the criticisms of Wien and Lummer as he would explain much later (Planck 1906, pp. 217–220). As we have seen above, in August 1900, Wien argued that the independence of the resonator contradicted somewhat the requirement of more than one resonator. Lummer pursued a similar idea, doubting that extensivity could be valid for a system of resonators. At last, Planck accepted these points: there must be some sort of interdependence between the resonators entailing that to calculate the change of entropy for the entire system, one has to know the change of entropy as a function of the average energy of a resonator. This consideration may have suggested to Planck that more specific information was needed on the average resonator energy U_ν, and such information could be extracted from the energy distribution among the resonators.

Apart from these critical points, there is another vague hint about a possible way to justify the new radiation law. On commenting the second derivative of the entropy, Planck states that "it is by far the simplest of all expressions which lead to [the entropy] as a logarithmic function of [the energy]—which is suggested from probability considerations" (Planck 1958, Vol. I, p. 689). Again, it is difficult to figure out what was in Planck's mind. When the logarithmic dependence between entropy and energy was established in the fourth paper of the Pentalogy, Planck did not allude to any connection with probability theory. In the Pentalogy, the logarithmic dependence was postulated to get the exponential factor of Wien's law. The March paper had slightly improved the situation because the exponential factor came out of an integration of a function of the type $1/U$. In that case, no logarithmic dependence was assumed. The same procedure was however inapplicable here because the correct second derivative had turned out to be a combination of a function $1/U$ with a function $1/U^2$, and there was no thermodynamic argument yielding such a weird combination. The idea that a justification of the radiation law required the derivation of the exponential factor was still valid, though. The quotation above seems to suggest that, in October, Planck was exploring the way that had appeared obvious to many of his contemporaries: to view the exponential factor as a manifestation

---

[6]A very similar law had been proposed in June by Lord Rayleigh (see Sect. 4.7.2).

[7]For a discussion of this passage see also (Kuhn 1978, pp. 96–97).

of same sort of law of errors. Both Michelson and Wien implicitly made this move when they related the exponential factor to Maxwell's distribution. Now, Planck was similarly evaluating the possibility of incorporating probabilistic arguments within his program.

4.5 The Quantum

It is safe to assume that Planck's acceptance of probabilistic arguments was not unconditional surrender. After all, he had fiercely opposed Boltzmann's statistical view of irreversibility for years. Presumably, his first step was to examine meticulously the formal properties of the new radiation law and the related entropy. To do that, he had to single out the contribution of the average energy of the resonator. Combining the results of the electromagnetic part of his theory, one can derive:

$$u_\nu = \frac{8\pi \nu^2}{c^3} U_\nu. \tag{4.3}$$

Working backwards from the correct distribution function, Planck could have easily found that (Rosenfeld 1936):

$$U_\nu = \frac{h\nu}{e^{h\nu/kT} - 1},$$

where h, k are two universal constants. If one derives $1/T$ from this formula and makes use of the thermodynamic identity $\frac{1}{T} = \frac{dS}{dU}$, one integration provides an entropy formula that can be arranged as follows:

$$S_\nu = k \ln \frac{\left(1 + \frac{U}{h\nu}\right)^{\left(1 + \frac{U}{h\nu}\right)}}{\left(\frac{U}{h\nu}\right)^{\left(\frac{U}{h\nu}\right)}} \simeq k \ln \frac{\left(1 + \frac{U}{h\nu}\right)!}{\left(\frac{U}{h\nu}\right)!}, \tag{4.4}$$

where the Stirling approximation has been used 'in reverse', from the exponential to the factorials. This formula has a combinatorial look by virtue of the factorials. In fact, Leon Rosenfeld claimed that the similarity between this formula and Boltzmann's definition of entropy ($S = k \log W$, where W is the state probability) induced Planck to adopt the combinatorial approach. This inference is a bit too quick, however. First, at that time, there was no such thing as the formula $S = k \log W$. It was Planck, after the fact, who wrote the entropy in such a general way. Second, it is not obvious to figure out a combinatorial model that encompasses the Eq. (4.4).[8] Some further reflection was needed.

---

[8]Some years later, for example, Hendrik Antoon Lorentz proved that the radiation law could be derived by a combinatorial model different from Planck's (Lorentz 1910).

If the Eq. (4.4) is a combinatorial formula, the ratio $U/h\nu$ must be an integer and a pure number. Exploiting these constraints, the ratio may be written as P/N, where P, N are both integers. Substituting in the Eq. (4.4) gives:

$$NS = k[(N+P)\ln(N+P) - N\ln N - P\ln P] = k\log\frac{(N+P)!}{N!P!}. \quad (4.5)$$

This manipulation provides an interpretation for N, because here Planck can exploit the extensivity property of entropy already deployed in March: N can be interpreted as the number of resonators. After all, it was true that one must use more than one resonator. Accordingly, $P = NU/h\nu$ became a number of 'energy elements', as P must be a pure quantity. This interpretation relies only on the formal properties of the Eq. (4.4), together with some general constraints on the combinatorial and the entropy. The number of energy elements is derived from the ratio between the other quantities. An echo of this procedure can be heard in the December paper, when Planck stated that "if the ratio thus calculated is not an integer, we take for P an integer in the neighborhood" (Planck 1958, Vol. I, p. 701). In the form (4.5), the entropy resembles patently a combinatorial formula. Now, the following step is to find a combinatorial model for this formula. In Boltzmann's writings there is the expression $\frac{(N+P-1)!}{(N-1)!P!}$, which is equivalent to the expression above if N and P are large numbers. It was likely this formal similarity, rather than the entropy formula, that led Planck to adopt Boltzmann's combinatorics. But this adoption still needs to be carried out carefully. To understand why, we have to pause for a moment and analyze Boltzmann's 1877 combinatorial models.

4.5.1 Boltzmann's Urn Model(s)

Two things are often forgotten about Boltzmann's use of combinatorics. The first is that the famous 1877 paper is not the only place where Boltzmann applied combinatorial techniques to arrive at the distribution law. He had followed the same strategy in a fundamental work of 1868, and more importantly, the formalism developed in that paper is key for understanding the 1877 combinatorial procedure (Boltzmann 1868, 1909, Vol. I, pp. 49–96).[9] The second thing is that there are three different combinatorial models in the 1877 paper. Let us start with the most famous one.

Boltzmann imagines that the total energy of the system can be divided into a finite number of equal elements $E = P\varepsilon$, such that each molecule can assume the energy $0, \varepsilon, 2\varepsilon, \ldots, p\varepsilon$, with P maximum energy allowed. The energy elements do not have a fixed magnitude.[10] A state is described by specifying its energy distribution, i.e., how many molecules have a certain number of elements. Boltzmann's aim is to calculate the probability of such a state. Given a certain distribution, there

---

[9]For a discussion of the 1868 paper see (Badino 2009, 2011).

[10]They must be small enough that their summation can be approximated by an integral, but they cannot be infinitesimal (Darrigol 1988; Hoyer 1980).

are many allocations of energy over individual molecules, which Boltzmann calls 'complexions',[11] corresponding to it. Specifically, all the complexions obtained by permuting a given one yield the same distribution, because the number of molecules with a certain energy is not changed when we swap two molecules. From combinatorics we know that if $D = [n_1, \ldots n_P]$ is a distribution in which n_i is the number of molecules with energy $i\varepsilon$ (or, equivalently, the number of molecules in the energy cell $i\varepsilon$), the number of possible complexions consistent with this distribution is:

$$C(D) = \frac{N!}{\prod n_i!}. \tag{4.6}$$

Counting the complexions corresponding to a state and dividing them up by the total number of possible complexions gives the probability of that state. Boltzmann proves that the equilibrium distribution is the most probable one, i.e., the one represented by the largest number of complexions.

To illustrate how this formalism works, Boltzmann conceives an urn model. Let us imagine a very large urn containing very many tickets on which the numbers $0, 1, \ldots, P$ are written. One ticket is drawn, a corresponding number of energy elements is assigned to the first molecule, the ticket is reintroduced in the urn, and the process goes on for N drawings. A set of N drawings gives a complexion. When such a complexion is obtained, two things may happen: either it fulfills the constraint on the total energy or it does not. This means that the sum total of the energy elements associated with the molecules must be exactly equal to the total energy of the system. Only in this case the complexion represents a physically possible state. In the other case, the complexion must be discarded. We keep drawing, checking, and selecting until, after a huge number of drawings, we have a large quantity of acceptable complexions. If the number of drawings is sufficiently large, the number of complexions corresponding to each distribution will be given by the Eq. (4.6). It is important to notice that the drawings concern groups of elements, which are ascribed to the molecule as a whole. In a generic drawing, an energy $i\varepsilon$, not i times the energy ε, is assigned to a molecule. From the point of view of the combinatorial model, this is a relevant difference. For the energy elements are simply understood as labels to distinguish different energy assignations. Hence, in this urn model, both the molecules and the energy allocations are 'distinguishable'. The model represents the allocation of distinguishable objects over distinguishable boxes.

For the calculation of the normalization factor, namely the total number of possible complexions, Boltzmann refers implicitly to his work of 1868. In that paper, he calculates a general expression for the total number of ways of distributing N objects over A possible boxes with a combinatorial constraint analogous to the energy conservation.[12] This number is $J(N, A) = \frac{(N+A-2)!}{(N-1)!((A-1)!}$. In the urn model discussed

---

[11]In the literature on combinatorics, a complexion (or variation) is a combination followed by a permutation or, in other words, all the combinations that can be obtained from an original one by permutations (Netto 1901).

[12]For the technical details see (Badino 2009; Costantini et al. 1996).

above, the number of possible values of the energy is $A = P + 1$, therefore the normalization factor becomes (Boltzmann 1909, Vol. II, p. 181):

$$J(N, P) = \frac{(N + P - 1)!}{(N - 1)! P!}. \tag{4.7}$$

This is precisely the formula Planck arrived at in his analysis of the entropy function. But if we go on reading Boltzmann's paper, which is something historians have seldom done, we realize that the model just described is not the only possible interpretation of this formula. In a further attempt at clarifying the issue, Boltzmann proposes a second urn model that is consistent with the definition of probability but is unacceptable because it does not fulfill the requirement of exchangeability (Boltzmann 1909, Vol. II, p. 172).[13] Toward the end of the paper, however, he describes a third urn model, which is most important for our story (Boltzmann 1909, Vol. II, pp. 211–214). Boltzmann imagines an urn containing molecules numbered from 1 to N. One carries out a set of P drawings reintroducing each time the molecule into the urn. At the end of this set of P drawings, a complexion is arranged by assigning to each molecule a number of elements equal to the number of times it has been drawn. In this model, there is no need to cancel out complexions because the total energy constraint is automatically fulfilled by drawing the molecules P times.

The key point is that the total number of complexions, in this case, is combinatorially known as the number of combinations of N elements of the Pth class and it is given again by the Eq. (4.7) (Netto 1901, pp. 20–21). The two combinatorial models lead to the same formula, but they are very different indeed. Contrary to the first model, here the molecules are numbered and are therefore distinguishable, but the energy elements are individually ascribed to the molecules and are indistinguishable. If a molecule has energy $i\varepsilon$, it has been individually drawn i times and each time it has received one element. However, a permutation of the energy elements does not change the complexion. Hence, the same formalism is compatible with two radically different combinatorial models. This is the most important point to bear in mind during an evaluation of Planck's use of combinatorics.

4.5.2 Planck's Adaption of the Combinatorial Technique

For many years, historians of physics debated one central problem: did Planck introduce the quantization of the energy of the resonator in his December paper or did he not? In this debate, Planck's use of combinatorial arguments has played a central role.[14] Martin Klein claimed that Planck did, and his main argument relied on the combinatorial model used by Planck (see Jost 1995; Klein 1962, 1979). Let us take a closer look at this model.

---

[13] See (Costantini et al. 1996).

[14] For a discussion of this debate see (Badino 2009; Darrigol 2001; Galison 1981).

Planck summarizes two months of intensive work in a short communication presented on 14 December 1900 at the German Physical Society. He assumes a system of free radiation and resonators at various frequencies. There are N_1 resonators operating at a frequency ν_1, N_2 at frequency ν_2, and so on. The N_i's are large numbers. The total energy of the system $E_s = E_r + E_N$ is the sum of the energy E_r of the radiation and the energy E_N allocated over the resonators. A macroscopic state is given by the distribution of energy over the different species of resonators: $D = [E_1, E_2, \ldots]$, where E_i is the energy ascribed to all the resonators with frequency ν_i.

What is the probability of D? Resonators of different frequencies cannot exchange energy, therefore the elementary states constituting the macroscopic distribution are the energy allocations over resonators of the same frequency. Thus, Planck defines the probability of D as $P(D) = \prod R_i$, where R_i is the number of ways of distributing the energy E_i over the N_i resonators at frequency ν_i and the product is extended over all species of resonators. The difference between this procedure and Boltzmann's is apparent. For Boltzmann, the probability is proportional to the number of microstates consistent with the distribution, while Planck considers the total number of permutations. Planck operates a selective adaption of Boltzmann's formalism. He knows that the model must contain the formula (4.7), which for Boltzmann represents the total number of microstates. To construe the model, he exploits the fact that the resonators are characterized by energy and frequency; thus, they require a further level of description, which is not present in kinetic theory where the molecules are described by energy only.

The energy to be distributed over the resonators at frequency ν_i is divided into elements $E_i = P_i \varepsilon_i$, where $\varepsilon_i = h\nu_i$ and h is a universal constant. Contrary to Boltzmann, the energy elements do have a fixed magnitude.[15] This constraint derives from the form of the distribution law. Then Planck states: "it is clear that the distribution of $[P_i]$ energy elements over $[N_i]$ resonators can only take place in a finite, well defined number of ways" (Planck 1958, Vol. I, p. 701). The total number of these distributions is:

$$R_i = \frac{(N_i + P_i - 1)!}{(N_i - 1)! P_i!}, \qquad (4.8)$$

in perfect analogy with the Eq. (4.7). From this statement Klein concluded that Planck thought of the resonator energy as quantized into discrete elements. Thomas Kuhn countered this reconstruction on the grounds that Planck's formalism is compatible with a more 'Boltzmannesque' combinatorial model.[16] In particular, Kuhn argued that Planck was counting the ways of distributing the N_i resonators over energy cells $(0, \varepsilon)$, $(\varepsilon, 2\varepsilon)$, $\ldots$. Because the resonators can be placed everywhere in the cell, there is no real quantization of the energy: the discretization of the energy continuum is

---

[15]In addition, they depend on the frequency, a circumstance that makes the calculation of the normalization factor very complicated. For this reason, Planck leaves it out entirely.

[16](Darrigol 1988; Kuhn 1978). As he was writing his dissertation under Klein's supervision, Allan Needell exchanged some letters with Kuhn precisely on this point. I thank Allan for sharing this important material with me.

only fictive. Kuhn's claim is not completely accurate though[17] and collides with Planck's overt statement in the December paper. For this reason, the debate never settled to a consensus.[18]

As a matter of fact, one can conclude very little about Planck's understanding of the resonator-field interaction from the statistical formalism he used. Planck resorted to Boltzmann combinatorics with a very specific goal: to construe a model able to support the combinatorial-like formula for the entropy. He had to adapt Boltzmann's approach to his particular case. To his wonder, he found two possible models in the 1877 paper. This ambivalence was perfectly functional to his epistemic trope. Combinatorics were yet another example of probabilistic techniques, i.e., symbolic practices to manipulate physically meaningful, albeit incomplete, information. For Planck, the use of such techniques meant automatically black-boxing the microscopic state. Again, the position expressed in the letter to Leo Graetz: we deploy probability because we do not have a microscopic account; if we had one, we would explain the phenomena in terms of micro-dynamics. The fact that the combinatorial formalism was compatible with two different microscopic accounts reinforced Planck's claim: probabilistic practice does not tell us something about the world, but rather about our cognitive status. His reading of the multiplicity of models was that the formalism did not require a commitment to any particular representation of the behavior of the resonator and the nature of radiation. The use of one model instead of the other is dictated by considerations of mere opportunity. In the first edition of the *Wärmestrahlung*, Planck explained this point very clearly. On comparing the combinatorial procedures for gas and for radiation, he pointed out that one may distribute the resonators over energy cells, but it is "faster and more convenient" to distribute individual energy elements over resonators (Planck 1906, pp. 151–152). The second way is clearly simpler because it automatically fulfills the energy constraint.

Thus, the ambivalence of the statistical formalism enabled Planck to maintain a noncommittal stance toward the adaption of Boltzmann's combinatorics. This lack of commitment was consistent with the spirit of his program. The introduction of the HNR had required the engineering of a conceptual space hinging on the distinction between the macroscopic quantities and the inaccessible realm of the microscopic processes.[19] Here, much like in the case of the HNR, a formal procedure fills a gap in the theory: the combinatorial technique leads to the conclusion of the argument without pointing at any specific microscopic behavior of the resonators. Thus, Planck can still look at the probabilistic practices as a convenient way to represent the unknown.

The second part of Planck's paper is a bit hasty and partly conceals another difference with Boltzmann's procedure. In Boltzmann's theory, the equilibrium distri-

---

[17]It is true, as I have shown above, that the Eq. (4.7) can be used to represent a distribution of energy elements over molecules or, alternatively, of molecules over energy cells, but the energy cells must be defined very carefully to allow the duality.

[18]For more information on this debate see (Darrigol 2000; Gearhart 2002; Jost 1995; Klein 1979; Kuhn 1984).

[19]On the centrality of this distinction in Planck's theory see also (Needell 1980).

bution is obtained by maximizing the state probability. This maximization, essential in kinetic theory, is superfluous in Planck's model. Understanding the reason is necessary for performing the calculations that Planck does not carry out explicitly in the original paper. The logarithm of the state probability is:

$$
\begin{aligned}
\log R &= \sum N_i \left[\left(1 + \frac{P_i}{N_i} \right) \log \left(1 + \frac{P_i}{N_i} \right) - \frac{P_i}{N_i} \log \frac{P_i}{N_i} \right] \\
&= \sum_i N_i \left[\left(1 + \frac{U_i}{h\nu_i} \right) \log \left(1 + \frac{U_i}{h\nu_i} \right) - \frac{U_i}{h\nu_i} \log \frac{U_i}{h\nu_i} \right].
\end{aligned}
$$

This expression should be maximized under the constraints that both the total number of resonators and the total energy remains constant. Deploying the usual technique of the Lagrangian multiplier, the maximization condition is:

$$
\delta \left(\log \mathbf{K} - \beta \sum_i N_i U_i \right) = 0,
$$

where β is an undetermined factor. Because the number of resonators is fixed, $\delta \sum_i N_i U_i = \sum_i N_i \delta U_i$, therefore:

$$
\delta \sum_i N_i \left[\left(1 + \frac{U_i}{h\nu_i} \right) \log \left(1 + \frac{U_i}{h\nu_i} \right) - \frac{U_i}{h\nu_i} \log \frac{U_i}{h\nu_i} - \beta U_i \right] = 0.
$$

Now comes an important point: by imposing the constraints above, we have made all the frequency-related quantities independent, thus it holds:

$$
\frac{\partial}{\partial U_i} \left[\left(1 + \frac{U_i}{h\nu_i} \right) \log \left(1 + \frac{U_i}{h\nu_i} \right) - \frac{U_i}{h\nu_i} \log \frac{U_i}{h\nu_i} - \beta U_i \right] = 0,
$$

for each ν_i. The problem is now a purely analytical one. By simple manipulation one gets:

$$
\frac{\partial}{\partial U_i} \left[\log(h\nu_i + U_i) + \frac{U_i}{h\nu_i} \log(h\nu_i + U_i) - \log h\nu_i - \frac{U_i}{h\nu_i} \log U_i - \beta U_i \right] = 0.
$$

The derivative is:

$$
\frac{1}{U + h\nu_i} + \frac{1}{h\nu_i} \log(h\nu_i + U_i) + \frac{U_i}{h\nu_i} \cdot \frac{1}{U_i + h\nu_i} - \frac{1}{h\nu_i} \log U_i - \beta = 0,
$$

which, after simplification, yields:

$$
\frac{1}{h\nu_i} \log \left(\frac{h\nu_i}{U_i} + 1 \right) = \beta.
$$

One can now immediately find the formula:

$$U_i = \frac{h\nu_i}{e^{\beta h\nu_i} - 1}.$$

Together with Eq. (4.3), this formula gives the correct radiation law. In the December paper, referring to the derivation of the radiation law from the entropy, Planck states that "[i]t would [...] be very complicated to perform explicitly the above-mentioned calculations", and therefore he resorts to "[a] more general calculation which is performed very simply" (Planck 1958, Vol. I, p. 703). Likely, he was referring to the fact that a precise calculation of the probability is extremely complex in his model because the frequency-dependence of the energy elements makes the evaluation of the normalization factor difficult. However, Planck must have realized very soon that the maximization procedure was not needed. In effect, the maximization only guarantees that the resonators are in thermal equilibrium, because the multiplier β turns out to be the usual exponent of the Boltzmann factor $1/kT$. But, as in the Pentalogy, the thermal equilibrium was assumed from the start. The form of the distribution function follows merely from the combinatorial formula (4.8), and it is superfluous to demand many species of resonators. For this reason, Planck drops this assumption in the *Annalen* paper published in January 1901.

4.6   The Quantum Hypothesis in 1900

The statement $\varepsilon = h\nu$ appears for the first time in December 1900 as part of Planck's statistical procedure. It plays an important role in this procedure—it allows the practical calculation of the entropy—but the procedure, in turn, is a scheme to interpret a still mysterious microscopic process. The incorporation of the combinatorial technique was a momentous step for Planck. To justify it, he felt the urge to relate the statistical notion of entropy with his previous work in radiation theory, and in so doing, he stressed the similarities between his approach and Boltzmann's more than ever. Disorder, he stated, was not a stranger in the Pentalogy:

> Entropy means disorder, and I thought that one should find this disorder in the irregularity with which even in a completely stationary radiation field the vibrations of the resonator change their amplitude and phase [...]. The constant energy of the stationary vibrating resonator can thus only be considered to be a time average, or, put differently, to be an instantaneous average of the energies of a large number of identical resonators which are in the same stationary radiation field, but far enough from one another not to influence each other directly. Since the entropy of a resonator is thus determined by the way in which the energy is distributed at one time over many resonators, I suspected that one should evaluate this quantity by introducing probability considerations into the electromagnetic theory of radiation. (Planck 1958, Vol. I, pp. 698–699)

Besides disorder, another essential ingredient of the combinatorial procedure was the equiprobability of the individual complexions. In this respect, Planck pointed out that this requirement "can be understood as a more detailed definition of the

hypothesis of natural radiation" (Planck 1958, Vol. I, p. 704). By making contact
between natural radiation and statistical arguments, Planck wanted to highlight that
they share the same conceptual space.

However, this conceptual organization of the theory disguises a fundamental frac-
ture. Planck's adoption of the Boltzmann's argumentative pattern in the Pentalogy is
based on an analogy between the gas molecules and the microscopic amplitudes and
phases. Remember the morphology described in Figs. 3.2 and 4.3: the connection
between the electromagnetic and the thermodynamic parts of his theory depends on
drawing the micro-macro divide in a way that amplitudes and phases belong to the
microworld, while intensity and average energy belong to the macroworld. Planck is
explicit on this analogy in the March paper, pointing out that it would be erroneous to
compare resonators and molecules. On adapting the combinatorial technique, however, the morphology of the theory changes dramatically (Fig. 4.5), and Planck relies
precisely upon an analogy between resonators and gas molecules. Just like molecules, resonators can be placed into energy cells or being the recipients of energy
elements. In Planck's combinatorial scheme, resonators behave like molecules did
in Boltzmann's. This shift in the analogical relation between resonators and molecules is extremely consequential. The microscopic state in the combinatorial part of
Planck's theory is a complexion, an allocation of energy over individual resonators.
But this is not a microscopic state in the radiation part: resonators are still a bundle of
rapidly varying amplitudes and phases. Planck's entire epistemic story hinges upon
the micro-macro divide, but this divide shifts from the electromagnetic and thermodynamic to the combinatorial part of the theory. This generates a deep conceptual
tension within Planck's theory, which would take many years to be unraveled: the
two parts can stay together only if we adopt a non-Boltzmannian way of counting
complexions.

There is an additional point to bear in mind to prepare for the discussion of the
next session. The combinatorial procedure that Planck decided to apply was by no
means a standard piece of kinetic theory. Boltzmann's combinatorial argument was
presented in 1877 with the goal of clarifying a conceptual point, the relations between
probability and the second law. Boltzmann himself did not commonly use that procedure to prove new results and other experts hardly mentioned it.[20] Thus, Planck
took up a peripheral symbolic practice of classical physics and made a radically new
deployment of it. In so doing, however, the links between Planck's theory and the
standard kinetic theory became uncertain and shaky.

Planck's attempt at showing continuity between the Pentalogy and the new combinatorial procedure stops before the quantum hypothesis. Figure 4.5 shows that the
morphology of Planck's theory has evolved into a two-pronged structure. The electromagnetic and thermodynamic parts comprise all the symbolic codifications and
practices we have seen in the previous section. These parts are still effective. It is
interesting to note, for instance, that Planck's second argument for irreversibility is

---

[20]There is only one exception: in 1884 Boltzmann used the combinatorial approach to handle the
problem of dissociation (Boltzmann 1884). On the reception of Boltzmann's combinatorial approach
see (Bryan 1894, pp. 91–95).

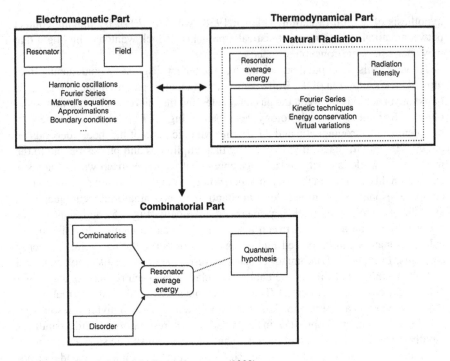

Fig. 4.5 Planck's theory in the December paper (1900)

still valid. The focus, however, has moved to the combinatorial part, which aims at establishing an expression for the average energy of the resonator. Here, Planck introduces new symbolic codifications and practices. What is more, these practices rely on an analogy that is at odds with the rest of the theory. But the most interesting aspect is that the combinatorial part requires a weird new item: the quantum hypothesis. On the one hand, this hypothesis prescribes a relation between the energy and the frequency of the resonator; therefore, it refers to the radiation part. On the other hand, it becomes necessary only in the statistical part. In addition, the hypothesis fixes the size of the energy element (or energy cell) by means of a universal constant, something that does not occur in Boltzmann's combinatorial procedure. Hence, the quantum hypothesis became an item 'epistemically isolated' from the rest of physical knowledge.[21] Far from attempting to connect this hypothesis with the radiation part, Planck took a non-committal stance toward the statistical procedure, thus separating the quantum from the description of the resonator behavior. Simultaneously, the combinatorial part, which offered the rationale for introducing the hypothesis, was separated from the original radiation theory and from Boltzmann's combinatorics. The epistemic structure of Planck's theory, constructed by progressively adapting foreign resources, could not find a place for the quantum. As a result, it was not even

---

[21]The concept of epistemic isolation for the interpretation of the quantum hypothesis has been developed in (Büttner et al. 2003).

clear whether the quantum hypothesis was a new symbolic practice to be used only in some problems or if it had an epistemic surplus telling us something about the world. Planck's contemporaries set out to debate precisely this point.

4.7 Reactions to Planck's Theory

In January 1901, Planck summarized the results of his short December communication, leaving out the issue of the maximization (Planck 1901, 1958, Vol. I, pp. 717–727). Unsurprisingly, Planck's combinatorial derivation stirred quite vigorous reactions, because if his radiation law was unquestionably well confirmed by the experiments, the theoretical justification was debatable to say the least. In this section, I divide the discussion on Planck's theory into two parts: the first deals with the critical examination of the radiation theory, while the second concerns the debate on the combinatorial procedure.

4.7.1 Electromagnetism in Context

On comparing Planck's program with his contemporaries in Sect. 2.3, I have pointed out that the Pentalogy purposely eschewed any specific assumption on the mechanism of exchange of energy between matter and radiation. What was considered a strong point by Planck was instead viewed as a fundamental flaw by all commentators. Hendrik Antoon Lorentz (1853–1928), the founder of electron theory, devoted some remarkable studies to the problem of heat radiation at the beginning of the twentieth century (Lorentz 1901a, b).[22] Lorentz's idea was the same as other experts such as Wien or Paul Drude: the black-body problem must provide a general and simple mechanism of interaction between matter and radiation that can be extended to more complex phenomena. This point is especially made in a paper on emission and absorption in metals. According to Drude's electromagnetic theory, absorption is due to charged electrons in complicated motion within the metals. From this presupposition, Lorentz drew a programmatic statement:

> Now we may infer from the relation between absorption and emission that is required by Kirchhoff's law, that the mechanism by which the emission of a body is produced is the same as that to which it owes its absorbing power. It is therefore natural to expect that, if we confine ourselves to the case of great wave-length, we shall be able to explain the emission of a metal by means of the heat-motion of its free electrons, without recurring to the hypothesis of 'vibrators' of some kind, producing waves of definite periods. (Lorentz 1902, p. 667)

This program is essentially different from Planck's because, "[a]s to the partition of energy between the vibrations of the resonators and the molecular motion in the body, Planck has not endeavored to give an idea of the processes by which it takes place"

---

[22]On the dissatisfaction of the contemporaries with Planck's theory see also (Garber 1976).

(Lorentz 1902, p. 668). Instead, Lorentz imagined that radiation emission follows the collision between electrons and ponderable matter because charges subject to sudden change of direction or velocity emit electromagnetic radiation. In this way, Lorentz had two important vantage points. First, he could use the Fourier series because radiation was regarded as the combination of irregular pulses stemming from the motion of the particles. Second, the probabilistic assumptions concerning the electronic behavior could be traced back to their original theoretical context, namely the kinetic theory of matter. Combining Drude's formula for absorption and his equation for emission derived by the same starting points, Lorentz could show that, in the limit of the long wavelengths, the radiation law depends linearly on the temperature. Although this result is less general than Planck's, Lorentz stressed the fact that it originated from completely different grounds.

Planck's theory was subject to critical scrutiny in order to understand its relation with irreversibility. Samuel Burbury, for instance, underscored the similarities between the argument in the fifth paper of the Pentalogy and the H-theorem. More importantly, Burbury was the first to realize that the key step to get Wien's law, the 'virtual' displacement of energy, contradicted the previous tenets of the theory: "[n]ow a 'virtual' change—if there is any virtue in the term—means a change consistent with the conditions of our system. In this supposition, therefore, we have definitely thrown overboard the restriction that energy can only be interchanged between systems of the same period" (Burbury 1902, p. 238).

Even deeper is the analysis published by Paul Ehrenfest (1880–1933) in 1905. Ehrenfest discussed the evolution of Planck's theory comparing the pre-quantum works with the papers published after 1900. Paralleling Planck's project with Boltzmann's kinetic theory, Ehrenfest asked "what are the—mutually independent—hypotheses that allow this theory to provide a cavity radiation with uniquely determined energy distribution for each temperature" (Ehrenfest 1905, p. 1301). Upon a close investigation, Ehrenfest discovered what had been already revealed by Burbury: contrary to Boltzmann's H-theorem, Planck's theory in its original form is not able to ensure the uniqueness of the stationary state. To prove this statement, Ehrenfest drew on a clever dimensional argument put forward by Lorentz (Lorentz 1901b, pp. 444–447).

Let us suppose in the cavity a radiation state Z_1 characterized by the natural radiation, a stationary value of the entropy S_1, total energy E_1, energy density Δ_1, and spectral distribution $s_1 = \phi(\lambda)$. Given the macroscopic state Z_1, each microscopic process going on in the cavity depends on Maxwell's equations and on the resonator equation. However, both are linear homogeneous equations of their arguments (electric and magnetic strengths, and dipole moment, respectively), therefore if we scale up these arguments by a factor m^2, we obtain a new radiation state Z_2 that has the same characteristics of the previous one (the macroscopic quantities are changed as follows $E_2 = m^4 E_1$, $\Delta_2 = m^4 \Delta_1$, $s_2 = m^4 \phi(\lambda)$). Now, this state has been obtained by working on the microscopic processes. But Planck's theory ascribes a unique spectral density as function of the macroscopic energy density only, namely as function of the temperature. Let us assume that this density is $s_2 = \psi(\lambda)$. Ehrenfest showed that it cannot be equal to the previous one because a violation of the displacement law would immediately follow. The conclusion is that the assumptions of natural radiation and stationary entropy value are not sufficient to ensure a unique spectral density. While in Boltzmann's theory the stationary value of the entropy corresponds unequivocally to Maxwell's distribution, in Planck's theory there is no general justification of the choice of the entropy function. As Ehrenfest would point out the following year (building on a Boltzmann's argument we have seen in Sect. 3.2.3), the resonators work as fixed obstacles for a stream of molecules: they diffuse them into the allowed space, but are unable to change the velocity distribution. Only molecular collisions can (Ehrenfest 1906).

Planck's theory reaches a stationary energy distribution because of a further assumption: the uniqueness of the entropy function. Ehrenfest argued that Planck's virtual process in the Pentalogy was equivalent to assuming the universal validity of his entropy. This analysis is undoubtedly correct, but it also reveals both the aims and the limitations of the commentators. For people with a strong background in statistical mechanics, such as Burbury and Ehrenfest, the comparison with Boltzmann's work was nearly the only way to make sense of Planck's theory. The irreversibility issue had to be recast in Boltzmannian terms to be understood and evaluated by the community. This process, however, ignited a misreading of Planck's program. As we have seen above, Planck never conceived the resonators system as a mean to change the energy distribution. Instead, resonators induce a modification in the spatial distribution of the field. The energetic equilibrium was assumed from the outset. Only after Boltzmann's criticisms, Planck was forced to rely heavily on the entropy concept to prove irreversibility. Succeeding adaptations of new conceptual resources shifted the focus of Planck's theory, which was not put forth as an attempt at deriving the radiation law. Thus, the reconfiguration of the theory in the context of statistical mechanics also dramatically changed its original goals and concentrated the attention upon the (misleading) resonators-molecules analogy.

4.7.2 Statistics in Context

The discussion of the statistical portion of Planck's theory was inspired by the different theoretical traditions within statistical mechanics. One of the most influential was the tradition concerned with the application of the equipartition theorem. One of Maxwell's most cherished results was the statement that energy is distributed equally over each degree of freedom. If a molecule can translate, vibrate, and rotate, each kind of motion obtains the same share of the total energy. This statement is known as equipartition theorem. Toward the end of the century, when kinetic theory was generalized to a state of matter more complex than gases, the discrepancies between the implications of the equipartition theorem and the experimental data, for instance on specific heats, were taken very seriously. The British community, especially G. H. Bryan, Samuel Burbury, Lord Rayleigh, William Thomson, and James Jeans, was particularly engaged in the attempt at understanding the role of the theorem in the interaction between matter and radiation. The problem of heat radiation was an ideal case for this research.

In June 1900, Lord Rayleigh published a short paper in which he drew some consequences about the radiation law from the point of view of the equipartition theorem (Rayleigh 1900). He pointed out that the form of Wien's law was not compatible with the theorem because it entailed that the energy tends to a finite value instead of increasing steadily as the temperature increases. Thus, Rayleigh proposed a new radiation law that contained both the exponential term and the direct dependence on the temperature. The rationale of this new law came from acoustic. Rayleigh argued that the number of normal modes of vibrations in a volume between the frequencies v and $v + dv$ is proportional to $v^2 dv$, hence the energy allocated in this region must be proportional to $T v^2 dv$. This apparently straightforward argument had a catastrophic implication of which Rayleigh was immediately aware. Because radiation occurs in a continuous medium (ether), the number of degrees of freedom is infinite; thus, the energy integral would diverge when calculated over all frequencies. To compensate for this consequence, Rayleigh combined the equipartition term $T v^2 dv$ with Wien's exponential term without giving any justification.

Around the same time, James Jeans (1877–1946) tackled the issue of heat radiation using the tools of analytical mechanics. His understanding of equipartition hinged upon a central idea: its generality notwithstanding, the theorem had some conditions of application, and the interaction between ether and matter was precisely the case in which these conditions broke down. Initially, Jeans relied on the fact that the smallest interaction between matter and ether transformed the problem into the analysis of a system with infinite degrees of freedom, for which no steady state can ever be reached (Jeans 1901). In the first edition of the *Dynamical Theory of Gases*, he further refined his argument. Because of the infinite number of degrees of freedom, there is no possibility to figure out a stationary state with equipartition of the available energy over the degrees of freedom. Jeans' way to make sense of the ether-matter interaction was to understand the exchange of energy as a dissipative phenomenon. During collisions, most of the energy remains translational, but a small portion is converted into molecular vibration, and from there, dissipated into ether. The process of dissipation, in turn, takes an extraordinarily long time if the molecular collisions are sufficiently brief: "we conclude that the rate of dissipation of the energy of a gas may reasonably be expected to be slow, provided that the product of the time occupied by a collision and the frequency of vibration is small" (Jeans 1904, p. 204).

Rayleigh did not understand Jeans' argument based on a dissipative exchange. For him, the problem was essentially like in acoustics: the distribution of energy over normal modes of vibrations. The idea that equipartition should establish itself rapidly among translations, but very slowly among vibrations, sounded like an illegitimate distinction between forms of mechanical motions (Rayleigh 1905a, b). In the case of heat radiation, the problem was particularly thorny. In 1905, when the debate between Rayleigh and Jeans took place, Planck's radiation law was already established experimentally beyond doubt. That law was however impossible to obtain by applying the equipartition theorem. A straightforward procedure gave a radiation intensity proportional to $RT\lambda^{-4}$, where R is the universal gas constant (Rayleigh

1905b, p. 55).[23] Experimental data were patently in disagreement with this formula, and this was enough, for Rayleigh, to conclude that "the law of equipartition cannot apply in its integrity."

In his reply, Jeans reiterated his argument and stressed that the slowness in the matter-ether exchange was not 'postulated' as Rayleigh thought, but followed from the application of those very tools of analytical mechanics on which Boltzmann had erected his kinetic theory (Jeans 1905b). For Jeans, his analysis suggested that the equilibrium between matter and ether does not occur within an experimentally measurable time, and therefore the equipartition theorem simply does not apply to such a case. However, when confronted with the classical analytical machinery, the statistical part of Planck's theory revealed hidden tensions. The fundamental assumption of Planck's statistical procedure, Jeans noticed, was that the probability for a resonator to have an energy between E and $E + dE$ depends on dE only. Now, the application of the Hamiltonian formalism to the behavior of a system in the phase space tells us that this assumption does not hold true over time in general (Jeans 1904, pp. 179–180). In the most general case, the system will tend to concentrate into certain energy cells. Planck's equiprobability of complexions is therefore ill-defined: "Prof. Planck's position is as though he had attempted to calculate the probability that a tree should be between six and seven feet high taking as his basis of calculation an enclosure of growing trees, and assuming the probability to be a function only of the quantities six and seven feet" (Jeans 1905a, p. 293).

The work of Rayleigh and Jeans was instrumental in reconfiguring Planck's statistical theory in the framework of the debate of equipartition theory and the relation between statistics and analytical mechanics. Although it was clear that their analysis and Planck's "are in effect the methods of statistical mechanics and of the theorem of equipartition of energy" (Jeans 1905a, p. 294), they apparently led to completely different results.[24] Rayleigh asked in despair "how another process, also based upon Boltzmann's ideas, can lead to a different result" (Rayleigh 1905b, p. 55).

Neither Rayleigh nor Jeans, however, explored the consequences for combinatorics involved in the resonator-molecule analogy. This investigation was carried out by Einstein and Ehrenfest. We have seen that Planck's peculiar attitude toward statistical arguments and, in general, the resources concerning the microscopic world, tolerated the status of 'epistemic isolation' of the quantum hypothesis. For Ehrenfest and Einstein, who were working within a strong Boltzmannian tradition, this isolation was much less tolerable. At the same time, both Ehrenfest and Einstein were aware that Planck had not used a standard piece of statistical mechanics. The fact that he had deployed it to solve such a deep riddle as the black-body radiation law added to the mystery. To clarify the role of Planck's combinatorics, it was necessary to see it as part of the wider application of statistics to physics.

---

[23] Initially, Rayleigh obtained an erroneous proportionality factor of 64π. The mistake was corrected by Jeans, who showed that, in accordance with Planck's Eq. (4.3), this factor is 8π (Jeans 1905b, pp. 97–98).

[24] On Jeans's subsequent conversion to quantum theory see (Gorham 1991).

Recent documents have shown that Albert Einstein (1879–1955) started this process of reinterpretation of Planck's statistics as early as 1901.[25] While Rayleigh and Jeans had been studying the features of the system molecules + radiation, Einstein figured out models in which the available knowledge on kinetic theory and radiation theory could be intertwined in order to reveal the hidden tensions. In 1905, Einstein imagined a cavity full of gas molecules, radiation, and electromagnetic resonators (Einstein 1905). On this system, one can apply two different arguments for the calculation of thermal equilibrium. If the resonators are in equilibrium with the molecules, their average energy must be the equipartition value; the equilibrium with the radiation, instead, corresponds to Planck's Eq. (4.3). Combining these two equations leads immediately to the Rayleigh-Jeans divergence of the energy integral. Einstein's simple argument has the merit of showing that the established knowledge of electromagnetism and mechanics, applied to the common concept of average energy, was incompatible with the experimental data. The statistical part, in other words, had to contain something that was, somehow, non-classical.

This new ingredient was the quantum hypothesis, a consequence of the application of the statistical arguments. Using the Boltzmann principle $S = k \log W$ 'in reverse', namely as a definition of probability, Einstein proved that the application of this argument both to the gas and to the radiation entailed a particle-like structure of radiation itself. Evidently, Einstein was taking the incorporation of Boltzmann's approach into radiation theory much more seriously than Planck himself. The latter had tried to adapt the combinatorial techniques in order to solve a problem; the former thought that the application of statistics to radiation theory should go hand in hand with precise ontological commitments about the microworld. In other words, Einstein was exploring the epistemic surplus of Planck's combinatorial procedure. Ehrenfest agreed. He argued that Planck's combinatorial way to irreversibility relied on two assumptions: (1) the radiation equilibrium follows from the most probable energy distribution over the resonators, and (2) the energy is divided into frequency-dependent elements. The second requirement, he admitted, had no analogon in Boltzmann's theory (Ehrenfest 1905, p. 1313).[26] Thus, dragging the quantum hypothesis out of the epistemically waterproof space conceived by Planck and re-contextualizing it in the available physical knowledge led progressively to the change of status of the hypothesis itself.

---

[25] See in particular the letters to Mileva Maric (Einstein 1987, Docs93, 97, 102, 111) in which Einstein criticized Planck's resonators and toyed with the possibility of applying Planck's law to specific heats. In my discussion of Einstein's analysis of the quantum, I follow closely (Büttner et al. 2003). For a broader discussion of Einstein's early investigations on statistical mechanics see (Renn 1997).

[26] See also (Ehrenfest 1906).

4.8 Conclusion

In this final section, I pull together the numerous strands of this long analysis and give some thoughts to the notion of 'quantum revolution'. It is tempting to call that fateful 14 December 1900 the birthday of a scientific revolution as though such an event can come about in a *fiat*. As an historical-epistemological process, however, a revolution must necessarily be more extended than the eureka moment. The foregoing analysis has shown that the revolution occurred as the consequence of the continuous change of the morphology of Planck's theory. This process, in turn, is due to the epistemic cooperation between the representational, transformational, and explanatory dimension.

The heat radiation as a borderline problem between different disciplines was the point of proliferation of several programs appealing to different traditions, which were characterized by different questions and fueled by different epistemic tropes. Planck's program was more rigid than others by virtue of his epistemic story based on a strict view of irreversibility, but it also adopted more flexible techniques to handle the radiation. That was necessary to avoid hypotheses on the microstructures. His first theory was articulated through a simple morphology in which the argument for irreversibility constituted the reasonable consequence of a well-defined electromagnetic problem.

At some point, this program had to face Boltzmann's criticism: this event can be considered the trigger of the overall revolutionary process. Boltzmann forced Planck to change the structure of his irreversibility argument with two prime consequences. In the first place, Planck had to reorganize his approach in order to make room for a new argumentative pattern. The peculiar combination of flexibility and rigidity sustained the adaption of kinetic theory. While other programs, such as Wien's, had used kinetic theory merely as a ready-made source of microscopic models, the formal structure of Planck's program allowed a deeper interaction that eventually unearthed the tensions between electromagnetism, thermodynamics, and kinetic theory. In the second place, the morphology of Planck's theory became more and more complex as the connection between its parts transformed. He introduced the HNR to frame the second argument and justified the leap from the electromagnetic to the thermodynamic part by appealing to physical meaninglessness of amplitudes and phases. In this way, Planck's program grew more and more entrenched with the fate of Wien's law. This amounted to an increase of structural rigidity that was further enhanced in March when Planck committed himself with a very specific form of the second derivative of the entropy. When Rubens revealed the outcomes of his experiments, Planck could not ignore them as he had done with Lummer's and Pringsheim's because the second derivative of the entropy was completely different. But interestingly, even when the theory had become so conceptually rigid to not withstand the experimental anomalies, the fundamental ingredients of Planck's program yielded a quick reorganization of the knowledge coming from the reservoir of Boltzmann's kinetic theory. It was at this point that Planck added to his program a further element: the combinatorial methods.

The December paper, however, further increased the tensions. Planck took another ingredient from Boltzmann's statistical mechanics: the combinatorial methods. He read them according to his view of the representational content of probabilistic tools: they indicate the inaccessibility of the microscopic state and entail that we had better black-box it and focus on the manipulation of average quantities. But when Planck put together the new combinatorial part and his previous theory, he shifted the micro/macro divide on which his construction hinged. The two parts could not stay consistently together, and they also required the weird quantum hypothesis.

This discussion teaches us, I believe, two important lessons. Lesson number one: theories are shifting objects. Contrary to the philosophical wisdom I have reviewed in the first chapter, 'Planck's black-body theory' is not a well-bounded system of statements, fundamental laws, or models, but it keeps changing over an extended period of time. The ways in which it represents the phenomena, transforms symbolic codifications, and explains equilibrium interact in a complex manner. New elements enter the theory, others disappear or take a less relevant role, or change their function. More importantly, they are all historically situated. Only by keeping track of this continuous change of morphology can one speak of 'Planck's black-body theory'.

Lesson number two: revolutions are at least as extended in time as the changes of theory morphology. It took a long, meticulous work to untie the bundle of conceptual tensions in Planck's 1900 theory. Boltzmann's legacy was kept alive by a new generation of masters and rising stars. For Lorentz, Ehrenfest, and Einstein, probabilistic methods encapsulated some epistemic surplus concerning the nature of radiation and the structure of the microstate. To understand Planck's theory and to make sense of the quantum hypothesis, they had to re-read it through the tradition of classical statistical physics they knew so well. Only when this tedious process of re-mapping Planck's theory in Boltzmannian terms was completed did the quantum gain an epistemic status, and it became clear that Planck had introduced something utterly new. Hence, the quantum breakthrough does not take place on 14 December, but it is an extended period of time from Planck's fourth paper to 1905, when physicists accept the fact that one cannot reconcile Planck's black-body theory with the rest of statistical physics. It is not a single—albeit momentous—concept that makes a revolution. Revolutions emerge retrospectively out of conflicting epistemic stories, representational contents, and mathematical practices, the tensional knots that they generate, and the explorative efforts necessary to untie these knots.

References

Badino M (2009) The odd couple: Boltzmann, Planck and the application of statistics to physics (1900–1913). Annalen der Physik 18(2–3):81–101

Badino M (2011) Mechanistic slumber vs. statistical insomnia: the early phase of Boltzmann's H-theorem (1868–1877). Eur Phys J H 36:353–378

Boltzmann L (1868) Studien über das Gleichgewicht der lebendigen Kraft zwischen bewegten materiellen Punkten. Sitzungsberichte der Akademie der Wissenschaften zu Wien 58:517–560

Boltzmann L (1884) Über das Arbeitsquantum, welches bei chemischen Verbindungen gewonnen werden kann. Annalen der Physik 22:39–72

Boltzmann L (1909) Wissenschaftliche Abhandlungen. Barth, Leipzig

Bryan GH (1894) Report on the present state of our knowledge of thermodynamics. Rep Brit Assoc Adv Sci 64:64–98

Burbury SH (1902) On irreversible processes and Planck's theory in relation thereto. Phil Mag 3(14):225–240

Büttner J, Renn J, Schemmel M (2003) Exploring the limits of classical physics: Planck, Einstein, and the structure of a scientific revolution. Stud Hist Philos Mod Phys 34:37–59

Costantini D, Garibaldi U, Penco MA (1996) Ludwig Boltzmann alla nascita della meccanica statistica. Statistica 3:279–300

Darrigol O (1988) Statistics and combinatorics in early quantum theory. Hist Stud Phys Sci 19:18–80

Darrigol O (2000) Continuities and discontinuities in Planck's Akt der Verzweiflung. Annalen der Physik 9:851–860

Darrigol O (2001) The historians' disagreements over the meaning of Planck's quantum. Centaurus 43:219–239

Ehrenfest P (1905) Über die physikalischen Voraussetzungen der Planck'schen Theorie der irreversiblen Strahlungsvorgänge. Sitzungsberichte der Akademie der Wissenschaften zu Wien 114:1301–1314

Ehrenfest P (1906) Zur Planckschen Strahlungstheorie. Physikalische Zeitschrift 7:528–532

Einstein A (1905) Über einen die Erzeugung und Verwandlung des Lichtes betreffenden heuristischen Gesichtspunkt. Annalen der Physik 17:132–148

Einstein A (1987) The early years, 1879–1909. The collected papers of Albert Einstein, vol 1. Princeton University Press, Princeton

Galison P (1981) Kuhn and the quantum controversy. Brit J Philos Sci 32:71–85

Garber E (1976) Some reactions to Planck's law. Stud Hist Philos Sci 7:89–126

Gearhart C (2002) Planck, the quantum, and the historians. Phys Perspect 4:170–215

Gorham G (1991) Planck's principle and Jeans's conversion. Stud Hist Philos Sci 22(3):471–497

Hettner G (1922) Die Bedeutung von Rubens Arbeiten für die Plancksche Strahlungsformel. Die Naturwissenschaften 10(48):1033–1038

Hoffmann D (2001) On the experimental context of Planck's foundation of quantum theory. Centaurus 43:240–259

Hoyer U (1980) Von Boltzmann zu Planck. Arch Hist Exact Sci 23:47–86

Jeans JH (1901) The distribution of molecular energy. Philos Trans Roy Soc London 196:397–430

Jeans JH (1904) The dynamical theory of gases. Cambridge University Press, Cambridge

Jeans JH (1905a) A comparison between two theories of radiation. Nature 72:293–294

Jeans JH (1905b) On the partition of energy between matter and ether. Phil Mag 10:91–98

Jost R (1995) Planck-Kritik des T. Kuhn. In: Das Märchen vom Elfenbeinernen Turm. Reden und Aufsätze, Springer, Berlin

Kangro H (1970) Vorgeschichte des Planckschen Strahlungsgesetzes. Steiner, Wiesbaden

Klein MJ (1962) Max Planck and the beginnings of the quantum theory. Arch Hist Exact Sci 1:459–479

Klein MJ, Shimony A, Pinch T (1979) Paradigm lost? A review symposium. Isis 70:429–440

Kuhn T (1978) Black-body theory and the quantum discontinuity, 1894–1912. Oxford University Press, Oxford

Kuhn T (1984) Revisiting Planck. Hist Stud Phys Sci 14:232–252

Lorentz HA (1901a) Boltzmann's and Wien's laws of radiation. Proceedings Koninklijke Akademie van Wetenschappen te Amsterdam 3:607–620

Lorentz HA (1901b) The theory of radiation and the second law of thermodynamics. Proceedings Koninklijke Akademie van Wetenschappen te Amsterdam 3:436–450

Lorentz HA (1902) On the emission and absorption by metals of rays of heat of great wavelengths. Proceedings Koninklijke Akademie van Wetenschappen te Amsterdam 5:666–685

Lorentz HA (1910) Alte und neue Fragen der Physik. Physikalische Zeitschrift 11:1234–1257

Lummer O (1900) Le rayonnement des corps noirs. In: Guillaume CE, Poincaré H (eds) Rapports presentes au Congres International de Physique, vol 2. Gauthier-Villars, Paris, pp 41–99

Lummer O, Pringsheim E (1899a) Die Vertheilung der Energie im Spectrum des schwarzen Körpers. Verhandlungen der Deutschen Physikalische Gesellschaft 1(1):23–41

Lummer O, Pringsheim E (1899b) Die Vertheilung der Energie im Spectrum des schwarzen Körpers und des blanken Platins. Verhandlungen der Deutschen Physikalische Gesellschaft 1(12):215–235

Lummer O, Pringsheim E (1900) Über die Strahlung des schwarzen Körpers für lange Wellen. Verhandlungen der Deutschen Physikalische Gesellschaft 2(12):163–180

Needell A (1980) Irreversibility and the failure of classical dynamics: max planck's work on the quantum theory, 1900–1915. PhD thesis, University of Michigan, Ann Arbor

Netto E (1901) Lehrbuch der Combinatorik. Teubner, Leipzig

Paschen F (1899) Über die Vertheilung der Energie in Spektrum des schwarzen Körpers bei höheren Temperaturen. Sitzungsberichte der Preussischen Akademie der Wissenschaften 2:959–976

Planck M (1900a) Entropie und Temperatur strahlender Wärme. Annalen der Physik 4:719–737

Planck M (1900b) Über eine Verbesserung der Wienschen Spektralgleichung. Verhandlungen der Deutschen Physikalische Gesellschaft 2:202–204

Planck M (1900c) Zur Theorie des Gesetzes der Energieverteilung im Normalspektrum. Verhandlungen der Deutschen Physikalische Gesellschaft 2:237–245

Planck M (1901) Über das Gesetz der Energieverteilung im Normalspektrum. Annalen der Physik 4:553–563

Planck M (1906) Vorlesungen über die Theorie der Wärmestrahlung. Barth, Leipzig

Planck M (1958) Physikalische Abhandlungen und Vorträge. Vieweg u, Sohn, Braunschweig

Rayleigh JWS (1900) Remarks upon the law of complete radiation. Phil Mag 49:539–540

Rayleigh JWS (1905a) The dinamical theory of gases. Nature 71:559

Rayleigh JWS (1905b) The dinamical theory of gases and of radiation. Nature 72:54–55

Renn J (1997) Einstein's controversy with Drude and the origin of statistical mechanics. Arch Hist Exact Sci 51:315–354

Rosenfeld L (1936) La premiere phase de l'evolution de la theorie des quanta. Osiris 2:148–196

Rubens H (1899) Über die Reststrahlen des Flussspathes. Annalen der Physik 69:576–588

Rubens H, Kurlbaum F (1900) Über die Emission langwelliger Wärmestrahlen durch den schwarzen Körper bei verschiedenen Temperaturen. Sitzungsberichte der Preussischen Akademie der Wissenschaften 2:929–941

Thiesen M (1900) Über das Gesetz der schwarzen Strahlung. Verhandlungen der Deutschen Physikalische Gesellschaft 2(5):65–70

Wien W (1900) Les lois théoriques du rayonnement. In: Guillaume CE, Poincaré H (eds) Rapports presentes au Congres International de Physique, vol 2. Gauthier-Villars, Paris, pp 23–40

Index

A
Analyzing resonator, 45, 63

B
Bartoli, A., 31
Bjerknes, V., 33
Boltzmann, L., 25, 26, 31, 32, 35–38, 48–53,
    58–60, 65–69, 71, 76, 82, 91–93, 98,
    99, 102, 103, 105–108
    combinatorial methods, 89, 92–96, 99,
        100, 108
    H-theorem, 49, 53, 60, 66, 68, 74, 102
    molecular chaos, 65, 67–69
Bryan, G.H., 66, 103
Burbury, S.P., 66, 102, 103

C
Carnap, R., 3, 5
Clausius, R., 37, 72
Complex morphology, 23, 24
    in Planck's theory, 58, 59, 76, 86, 90, 99,
        107, 108
Culverwell, E.P., 65–67

D
Drude, P., 101, 102

E
Ehrenfest, P., 102, 103, 105, 106, 108
Einstein, A., 106, 108
Epistemic cooperation, 21, 23, 24, 55, 58,
    63, 65, 77, 86, 107

Epistemic surplus, 21, 23, 57, 60, 61, 68, 76,
    89, 101, 106, 108
Epistemic trope, 22–24, 47, 60, 96, 107
Explanatory dimension, 20, 22, 23, 33, 39,
    46, 53, 58, 65, 70, 76, 107

F
Fourier series, 23, 47, 48, 54, 56, 57, 61–63,
    65, 72, 73, 76, 102

G
Gouy, L.G., 56, 62, 63, 70
Graetz, L., 45, 52, 60, 96

H
Helmholtz, H., 72
Hempel, C.G., 3, 5, 11
Hertz, H., 32, 36, 42–44, 52, 54, 55, 59, 61
Hypothesis of natural radiation, 23, 25, 57,
    61, 65, 68–72, 76, 77, 96, 99, 107

I
Irreversibility
    Planck's first argument, 46, 52, 53, 56,
        58, 59, 66, 75, 76
    Planck's second argument, 61, 71–73,
        76, 77
    Planck's view of i., 45, 47, 60, 75
    statistical view of i., 45, 50, 60, 62, 91

J
Jeans, J.H., 66, 69, 103–106

© The Author(s) 2015
M. Badino, *The Bumpy Road*, SpringerBriefs in History
of Science and Technology, DOI 10.1007/978-3-319-20031-6

Printed in the United States
By Bookmasters